绘心绘色

新意象 家居配色

ORIGINAL MINDS ANG VIVID
COLORS

NEW IMAGES OF
HOME COLOR
COLLOCATION

Shenzhen Design Vision Cultural Dissemination Co., Ltd
深圳视界文化传播有限公司 编

中国林业出版社
China Forestry Publishing House

色彩设计生活
COLOR DESIGNS LIFE

序言
PREFACE

研究表明，人进入一个空间，最开始几秒钟内获得的印象，75%是对色彩的感觉，之后才会去了解形体。色彩是室内装饰设计中不可以忽略的关键因素，也是室内设计中最具表现力和感染力的因素。

赛拉维，CEST LA VIE 法文译为：这就是生活。我们作为专业的室内设计师，不只做设计，更是美好生活场景的营造者。

在室内设计中，色彩运用的本质并不是色彩本身，而是作为营造生活场景的表达手段。通过有效地改善空间环境，对不同色彩的层次运用，来满足居住者的个性及生活习惯需要，同时传递出居住者的身份、个性、品位和情趣等，营造不同的生活场景，让居住者在通过室内环境勾勒下能做出对未来美好生活的想象和期待。

例如橙色调是激情、温暖的表达，适用于充满活力的多口之家；灰色调偏冷，时尚静谧，与金属搭配可提升空间质感，加上华丽高雅的家居配饰，营造现代感十足的家居氛围，更适合独自居住的都市时尚年轻客群。

色彩的多样性，使室内设计变得有趣且变幻无穷。色彩赋予相同的空间以不同的风格及层次，在室内设计中充分发挥色彩的特点，理性地运用色彩的感性倾向，创造出和谐多样的生活情境。同时，我们应该认识到室内设计中的色彩是手段不是目的，是在对客群的了解基础上运用设计师的独特艺术性加以表达，这样的设计作品不单是设计师的自我发挥，也是对业主的理解与演绎。总的来说，变的是色彩，不变的是设计师对品质生活的探索和追求，我想这应该也是设计师最想传达的设计理念。

Researches show that, when people enter into a space, the impression they have in the first few seconds is the feeling about colors which counters for 75% and then the understanding of forms. Color is a nonnegligible key factor in interior decorative design and is the most expressive and infectious factor in interior design.

CEST LA VIE means this is life in French. As professional interior designers, we not only do designs, but also create better life scenes.

In interior design, the essence of color application is not the color itself but the expression method to create life scene. Effectively improving the space environment and applying colors in different shades to meet residents' needs of personalities and living habits, convey their identities, personalities, tastes and interests and create different life scenes to make them imagine and expect better future lives through outlines of interior environment.

For example, orange is the expression of passion and warmth and is suitable in a vigorous family with many members. Gray is cold, fashionable and tranquil and can promote texture of the space when collocated with metal; in addition with gorgeous and elegant home furnishings, it can creates a modern living atmosphere which is more suitable for urban fashionable young clients to live in.

The diversity of color makes interior design interesting and changeable. Colors endow the same space with different styles and gradations. Making full use of features of color in interior design and rationally using perceptual tendencies of color can create harmonious and diversified live scenes. At the same time, we should realize that color in interior design is the method rather than the goal and is to use designers' unique artistries to express on the basis of understanding target clients so that this kind of design work is not only the designer's self-performance but the understanding and interpretation of owners. In general, what changes is color while what doesn't change is designers' exploration and pursuit of quality life. I think this might be the design concept which designers want to convey most.

王少青 / 赛拉维室内装饰设计（天津）有限公司
Shaoqing Wang/CLV–DESIGN

目录 CONTENTS

黄
YELLOW

008 热情洋溢 泰式大宅
EBULLIENT THAI STYLE MANSION

018 摩登洛可可 金色殿堂
MODERN ROCOCO, GOLD PALACE

026 温婉与灵秀之美
GENTLE AND EXQUISITE BEAUTY

046 流金溢彩
FLOWING AND SHINING GOLD COLOR

064 重现法式时光
REPRODUCING FRENCH TIME

072 夏日里的丛林世界
JUNGLE WORLD IN SUMMER

082 淡抹相宜 雅致相趣
CHARMING AS ALWAYS, ELEGANT AND FUNNY AS USUAL

蓝
BLUE

210 包容中式 幸福之家
INCLUSIVE CHINESE STYLE, HAPPY HOME

220 一次关于文化与信仰的探索
AN EXPLORATION ABOUT CULTURE AND FAITH

230 三代同乐 典雅居所
ELEGANT RESIDENCE WITH JOYS OF THREE GENERATIONS

238 邂逅法式风情
ENCOUNTERING FRENCH STYLE

橙
ORANGE

- **090** 东情西韵 悦享人生
  ENJOY LIFE WITH ORIENTAL AND WESTERN CHARMS
- **110** 英伦风情 品位经典
  BRITISH STYLE, CLASSIC TASTE
- **122** 摩登时尚 自在之家
  MODERN FASHION, COMFORTABLE HOME
- **130** 爱马仕时尚家
  HERMES FASHION HOME
- **136** 墨绿轻奏 红橙浅唱
  DARK GREEN PLAYS LIGHTLY, RED ORANGE SINGS GENTLY
- **142** 橙色魅力 至美之家
  CHARMING ORANGE, BEAUTIFUL HOME

红
RED

- **152** 现代法式 东方典藏
  MODERN FRENCH STYLE, ORIENTAL COLLECTION
- **164** 锦绣满堂
  SPLENDIDNESS
- **176** 绽放的英伦玫瑰
  BLOOMING BRITISH ROSE
- **188** 玛莎拉红 高贵灵韵
  MARSALA RED WITH NOBLE AURA
- **202** 品享生活
  ENJOYING LIFE

蓝

- **246** 浪漫爱琴海
  ROMANTIC AEGEAN SEA
- **256** 绅蓝空间
  GENTLE BLUE SPACE
- **262** 住在米兰的鬼马小姐
  MISS FAIRY IN MILAN
- **268** 海洋坐标去旅行
  TRAVEL ON THE OCEAN

灰
GRAY

- **276** 传承人文新风范
  INHERITING HUMANISTIC NEW STYLE
- **286** 春风笑 润绿珠
  SPRING BREEZE SMILES AND MOISTENS GREEN BEADS
- **298** 宁静港湾
  TRANQUIL HARBOR
- **304** 素念禅心
  PLAIN MIND AND ZEN HEART
- **312** 午后伯爵茶
  AFTERNOON EARL GRAY TEA

亮丽神圣
是你如太阳般的光彩
散发着柠檬与迎春花般的香气
高贵 快活
搭配紫蓝黑 积极辉煌
搭配橙与绿 相得益彰
阳光灿烂间骄傲威严

Y E L L O W

热情洋溢 泰式大宅
EBULLIENT THAI STYLE MANSION

>>> 设计理念 | DESIGN CONCEPT

If Chinese culture defines Zen and self-examination, then Southeast Asian culture defines leisure and passion. At the beginning of the design, ZESTART selectively deconstructs and reorganizes Thai style, strives to present a space atmosphere for residents to enjoy life and weakens the decorations, luxuries and vulgarness. The choices of many eye-catching colors and downplays collocation present the tropical feelings which are exactly the passion, happiness and cheers brought by tropical flavors to people. Being here, you will relieve tense emotions unconsciously, sit casually, put aside the bothering earthliness and forget multifarious things around you.

假如说中国文化定义了禅意与自省，那么东南亚文化则定义了休闲与热情。则灵艺术在设计之初，有选择地对泰式风格进行了解构与重组，力求展现居者享受生活的空间氛围，而淡化处理装饰、奢侈与脂粉气。多种醒目色彩的选择和细致的搭配呈现出热带的缤纷，正是那热带风情带给人们热情、欢愉和爽朗的感受。热带风情使人置身其中，不自觉地舒缓紧张情绪、随性坐卧、抛开纷纷扰扰的俗世，遗忘身边繁杂的琐事。

项目名称：昆明滇池龙岸云玺大宅十三区泰式户型别墅样板间
设计公司：深圳市则灵文化艺术有限公司
设计师：罗玉立
项目地点：云南昆明
项目面积：530m²
主要材料：实木、金属、石材、布艺等
摄影师：陈中

>>> 客厅：透天阳光

　　双层挑高的客厅设计别出心裁，天花以透明玻璃为顶，背景墙大面积使用象征着太阳光辉的黄色，阳光透过玻璃天花洒落在黄色墙面，自然和谐又温暖，夜晚在此足不出户，也能观赏良辰美景。温暖的黄色和热情的红色同时又相伴相生，沙发椅以红色为主，客厅中央的蓝色坐垫正面呼应格栅，高大的热带绿色植物充满了向上的生命力，空间的一切都彰显出热带风情的活力和魅力。

>>> 造景：隔不断的蓝色风情

　　蓝色的格栅增加空间的交流隔而不断，隐约间能观赏到客厅透过来的精彩。红色的桌几上摆放着充满泰式风情的插花和瓷器摆件，以蓝色为主色调色彩缤纷的坐椅，配上一片蓝的地砖更显空间魅力。各色半透明的吊灯五彩缤纷，当你走进来便会感染到空间传递给你的热情和欢乐。

主体色
PRIMARY COLOR

点缀色
INTERSPERSED COLOR

>>> 色彩分析 | COLOR ANALYSIS

This Thai style show flat in the thirteen district of Dianchi Dragon Shore Yunxi Mansion in Kunming sets yellow as the main tone at the same time uses gorgeous red, amorous orange, old coffee, dazzling blue and fresh green to create such a fascinating and charming and ebullient space, as if bringing the viewers to see the luxuriant rain forests and enjoying the vigorous plants, brilliant sunshine and unexpected heavy rains.

  昆明滇池龙岸云玺大宅十三区泰式户型别墅样板间以黄色为主打色彩，同时用绚丽的红、多情的橙、古老的咖、醒目的蓝、清新的绿共同搭配成这样一个风情万种而又热情洋溢的空间，仿佛带领参观者看见茂盛的雨林，正感受着勃勃生机的植物、灿烂的阳光和不期而至的暴雨。

>>> 会客厅：休闲里的时光

比起客厅的正式大气，会客厅则显得随意自由。同样带有热带风情的彩色背景墙缤纷耀眼，米白色和蓝色长条沙发上点缀着各色的抱枕，麻色的地毯和外露木色横梁则又为这个空间增添了些许乡村风格的休闲气息。

>>> 餐厅：淡雅醒目

相比热情缤纷的客厅，餐厅在色彩上处理偏向冷静和理智，这里更多的是带给就餐时的一份安静和淡雅。黄色依然是空间醒目的主角色，餐椅设计以稳重深沉的木色为轮廓搭配清新的薄荷绿坐垫，椅背以金色刺绣显示出做工的精致。浅木色的桌面上整齐摆放着摆件和餐具，黄色的餐巾和摆件，顿时点亮了就餐时的氛围。

>>> 主卧：缤纷的热情与欢乐

　　主卧看似无序实则有序地运用了多种色彩，尽管如此，我们还是能从中找出背景色的灰，主角色的黄，点缀色的红以及配角色的褐。各色的相撞传递出空间的欢乐和幸福之感，宽敞的落地窗又及时送来明媚的阳光，泰式风情的吊灯、摆件和墙面装饰，又赋予了空间东南亚的独特魅力。

摩登洛可可 金色殿堂
MODERN ROCOCO, GOLD PALACE

>>> 设计理念 | DESIGN CONCEPT

If baroque style dominates French classical public architectural style, then when it comes to residence, rococo style is more suitable to deduce aesthetics and comfort. How to translate the 18th century rococo style into modern space languages becomes the designer's major challenge.

Rococo style is good at using asymmetry to create aesthetics; the lines are delicate and euphemistical; the colors are tender and gorgeous; it inherits the vanity and magnificence of baroque style and endows solemn classical art with soft slenderness and complication. Among it, there flows a wakeful spirit of liberalism. The designer's interpretation of rococo is exactly the starting point of the design of this palace.

如果说巴洛克艺术主导了法国古典公共建筑的风格，到了宅邸内，洛可可风格更适合演绎唯美和舒适。如何把18世纪的洛可可艺术风格翻译成现代摩登的空间语言，成为设计师面临的主要挑战。

洛可可风格善用不对称营造美感，线条纤弱婉转、颜色娇嫩华丽，继承了巴洛克格调的浮华与盛大，又将庄严的古典主义艺术大胆赋予了柔美的纤细与繁复，其间流淌着一种自由主义觉醒的精神，设计师对洛可可的解读，正是这座宅邸设计的起点。

项目名称：北京格拉斯小镇
设计公司：IADC涞澳设计
设计师：潘及
项目地点：北京
项目面积：600m²
主要材料：大理石、布艺、铁艺等

>>> 客厅：古典气质与装饰艺术共生

　　这种复古而又时尚的感觉与 Art Deco 一同奠定了整个客厅的基调。金色穹顶下的一大一小两盏灵动多变的巴卡拉水晶吊灯也是不对称的设计，增加了空间丰富的层次感。

　　雕像底座上摆设的是饰萨尔瓦多·达利最钟爱的作品之一"爱丽丝梦游仙境"，绳子缠绕成一股连接着爱丽丝的手臂，爱丽丝的手和头发都幻化成象征着女性美丽芬芳的玫瑰花。爱丽丝混淆了现实与魔幻，在这个空间中则意味着混淆了洛可可风格的古典与摩登风格的现代：既有华丽轻快、精美纤细 18 世纪古董椅子，也有现代简约风格的沙发，两者的强烈对比被有着现代造型洛可可色系的茶几融合。此外，设计师还精选了 Horst P. Horst 等光影大师的摄影作品，进一步增加了室内的艺术气息。

>>> 餐厅：娇媚欲滴

洛可可的另一大特色就是颜色娇美，用金色及鲜艳色的浅色调打造奢华的氛围；并运用当时名贵稀有的东方纹样进行装饰。二者汇聚一起，娇美华丽。

>>> 门厅入口：水墨美学

进入大宅，地面上蜿蜒的 S 型水墨曲线像落入了一池，随波荡漾，通向大宅深处。古典的泼墨画般的地面像水中的舞者，轻盈地跳跃、旋转、翻腾，柔软的身姿，美丽的形态，合着水在一起流动，水流到哪里，美就延伸到哪去。

第一眼望到的就是走廊尽头的号称"德国国宝"的博兰斯勒古董钢琴，该品牌自问世以来就被欧洲众多国家皇室指定为收藏乐器，也是如鲁宾斯坦、亚历山大·帕雷、刘诗昆、周广仁等众多钢琴大师的首选。沉淀了厚重历史的黑色钢琴沐浴着铂金色的灯光，在象牙白的廊道终点熠熠生辉。不对称的空间诠释了洛可可的特点，散发着高贵和优雅的钢琴引入了摩登的感觉。

主体色
PRIMARY COLOR

点缀色
INTERSPERSED COLOR

>>> 色彩分析 | COLOR ANALYSIS

The designer uses ivory white as the main tone to create a rococo artistic space with historic connotations; color sense of the entire space is white, clean and transparent. At the same time, gold with aristocratic smell is chosen to deposit texture of the space, which is gorgeous and noble and manifests the status as a king. The skillfully interspersed red and green are lively and fiery and full of vitality. The entire mansion integrates strong palace flavors which were popular in Europe in 18th century with noble tastes of modern upper class; elegant radian and exquisite lines create classical charms by modern technique and materials. This is exactly the essence of modern rococo style. Delicate and charming, gorgeous and exquisite, it creates the warmth of home by typical feminine artistic style.

以象牙白为主色打造富有历史底蕴的洛可可艺术空间，整体色感白净、透明。同时选用极具贵族气息的金色沉淀空间质感，既华丽贵气又彰显王者地位。而巧妙点缀的红色、绿色等，活泼热烈，满满的生命力。整个宅邸融合了18世纪风靡欧洲的浓烈的宫廷气息与现代上流社会的高贵品味，优雅的弧度、精致的线条，以现代手法和材质缔造出古典神韵，这正是摩登洛可可的精髓所在：纤弱娇媚、华丽精巧，以典型的女性化的艺术风格打造家的温暖。

>>> 休闲空间：色彩明丽，曲线柔美，有节奏感

这套宅邸，不仅仅承载了住宿的功能，还是重要的社交场所。整个地下室空间就是为主人量身打造的休闲场所，以大都会 Pub 风格为背景，兼具酒窖、雪茄吧、台球室等功能。红色的吊顶与咖啡和白拼花的砖形成了强烈的视觉冲击；金色的隔断、灯饰与背景墙注入了奢华的气息，是实力的象征。整个空间让人流连忘返，惟愿沉溺其中，不可自拔。

>>> 卧房：色彩舒适与氛围活泼兼顾

鼠尾草绿的老人房跳脱了人们以往对于老人房刻板守旧的印象，充满活力但又不失稳重，仙鹤的纹样倾注了对老人最衷心的祝福。女孩房的芍药粉不仅娇嫩可爱，又有一丝贵族庄重的涵养；纯手工的LLADRO瓷偶，是欧洲上流社会每一个小女孩人手必备的玩具。

温婉与灵秀之美
GENTLE AND EXQUISITE BEAUTY

>>> 设计理念 ｜ DESIGN CONCEPT

Through surreal aesthetic creative technique, the designer integrates the vast ocean, vibrant earth and dreamlike aesthetic sky into this aesthetic and romantic modern French living space to bring you the strongest visual impact and to deduce surreal aesthetics originated from nature. White crystal droplight pours down along with the stair, which is glittering and translucent, luxurious and aesthetic. Flowers on behalf of nature and life are everywhere, sending out romantic and emotional soft breaths. Large pieces of window design not only bring visual transparency, but also bring more natural lights into interior, creating a dynamic living space. The designer applies classic elements such as modern fireplace and stone pillars into the design and freely interprets the enduring charm of elegant French style.

设计师通过超现实唯美的创意手法，将浩瀚无边的海洋、生机盎然的大地，抑或是梦幻唯美的天空全部融入到这个唯美浪漫的现代法式家居空间，带给你最强大的视觉冲击，演绎源于自然的超现实唯美。白色水晶吊灯顺应楼梯中空垂直而下，晶莹剔透，兼具奢华与美感。代表自然与生机的花卉随处可见，花开满室，花团锦簇，弥漫着浪漫而富有情调的温柔气息。大开窗的设计不仅带来视觉上的通透感，也将自然光线更多地容纳进来，打造充满活力的生活天地。设计师将现代壁炉、石柱等经典元素运用到设计中，收放自如地诠释了优雅法式经久不衰的魅力。

项目名称：绿城·临安青山湖红枫园
设计公司：浙江绿城家居发展有限公司
设计师：绿城家居设计团队
项目地点：浙江临安
项目面积：700m²
主要材料：胡桃木、金属、皮革、大理石等
摄影师：翰珑广告

主体色	
PRIMARY COLOR	

点缀色
INTERSPERSED COLOR

>>> 色彩分析 | COLOR ANALYSIS

The whole color collocation sets fashionable and warm keynote. The space gives priority to creamy beige, matching with senior ash and pearl white, which is elegant and tranquil and makes a harmonious and natural visual transition. Blue and red intersperse in the space, promoting vividness of the space and acting as striking points.

整体配色围绕着时尚、温暖的主基调依次展开。以米黄色为主，搭配高级灰、珍珠白，配色素雅、恬淡，视觉上过渡和谐自然。蓝色及红色点缀在空间中，提升了空间的鲜艳度，起到点睛作用。

>>> 客厅：如花安然静守

浅灰的护墙板辅之米色大理石地面，线条简洁流畅。造型典雅的壁炉在视觉和功能上都注入无可比拟的暖意。大型花卉画作，米黄交错的漩涡及曲线，充满梦幻。环形沙发呈现丝绒特有的光泽和柔顺质感，无形中彰显品位。

>>> 餐厅：众星拱月

座椅呈环形分布，如众星拱月般将红棕色实木餐桌环绕，在对比中强化视觉中心。明艳的迎春花束，十分吸睛。即使足不出户，也能感受到浓浓的春意。

>>> 主卧：恬淡素心

　　米色墙面由灰色线条勾框，赋予空间恬淡意味，亦能久处不厌。银灰色床品搭配蓝色床尾沙发，烘托出和谐舒适的居家氛围。沙发哑光的色泽感，让它多了一份无法细说的恬静。

>>> 客卧：纤尘不染而为清

亮白色背景搭配青色主题壁纸，冷静的色调带来优雅质感。菱形图案的床头背景搭配大小不一的金属装饰，特别的排列次序及精巧设计，使本来单调的背景墙有了活力。相同的色调属性令空间的基调和谐而统一，静谧而从容。

>>> 男孩房：和风容与

男孩子性格活泼好动，所以设计师采用菱形及V形的几何图案来装饰整个空间，灵动而富有变化。蓝灰条纹扶手凳与V形图案的地毯颜色呼应，给人活泼的视觉感。

流金溢彩
FLOWING AND SHINING GOLD COLOR

>>> 设计理念 | DESIGN CONCEPT

Positioned as Italian neo-classical style, this project has four layers. The first floor is reception and dining areas with living room, dining room and rest room. The second floor is bedrooms for sleeping with master bedroom and two boy's rooms in different styles. The underground two floors are mainly activities rooms for owners and guests; the original Italian style is injected with the owner's favorite Chinese elements with strong colors; furnishings and decorative paintings continue Chinese theme, Zen-like and eye-catching.

This project not only boldly uses colors and makes great breakthroughs in the foundation of traditional neo-classical style, but also combines with the owner's needs, which is also a light spot.

本案为意式新古典风格，上下共4层。一层为会客餐饮区域，分布客厅、餐厅及休息室等空间。二层主要为休憩空间的卧室，分布有主卧及两个风格迥异的男孩房。地下一层以及二层为业主与客人活动较多的区域，在原本意式的风格之上加入了业主喜欢的中国元素，色彩较为浓烈，同时又在饰品及装饰画的选择上延续中式主题，禅意且吸引眼球。

该案例不仅颜色运用大胆，在传统的新古典的风格之上做了很大的突破，而且结合业主的需求使用，也是一大亮点。

项目名称：绿城·台州明月
设计公司：浙江绿城家居发展有限公司
设计师：严晨、许嘉丹
项目地点：浙江台州
项目面积：1000m²
主要材料：石材、壁纸、软包等
摄影师：张静

>>> 色彩分析 | COLOR ANALYSIS

The whole space sets beige gold as the basic tone; the original shining gold is injected with soft beige to reconcile; the designers use modern and creative ways to make gold present an unprecedented charm. The designers use adjacent colors in different shades and choose pattern blends or fabric and texture blends so as to make spaces with gold elements softer.

整体空间以米金为基础色调，原本熠熠生辉的金色中加入柔和的米色调和，用现代又极具创意的方式令金色呈现出前所未有的魅力。为了让含有金色元素的空间看起来柔和些，设计师除了使用不同明度的邻近色，还选择图案混合或面料和质地的混合（软硬结合）来实现目的。

主体色 PRIMARY COLOR

点缀色 INTERSPERSED COLOR

>>> 客厅：大气与精致并存

一进客厅，就被其古典美感与奢华质感深深吸引。挑高空间开阔向上，金色灯带垂泻而下，共同营造盛大宏伟的空间气势。黄铜摆件、描金茶几、雕花椅凳在灯光的映射下，细节考究，打造金碧辉煌的效果。

>>> 餐厅：中西文化的融合

设计师结合业主崇尚欧洲的生活方式及喜爱中国传统文化的这一特点，在餐厅选择椅背带有黑金镶边的中式镂空雕花餐椅。除了用业主钟爱的酒架作为装饰背景，墙面一幅带有中式描绘元素的木质装饰镜，是整个餐区的亮点所在。

>>> 休息室：共度悠闲时光

家庭室两面大落地门窗，正对户外中庭水池和花园，景致优美。屋内家具选用孔雀绿暗花纹的精致绣花面料，搭配中式小边柜和画饰，使空间慵懒、妩媚又不失雅致，是女主人享受下午茶的好地方。

>>> 地下二层：永不凋零的经典

在黑色背景的衬托下，金色线条显得格外耀眼。而饱满浓郁的红色与金色结合亦是如此，这种组合方式成为了简单易行的设计经典。

>>> 主卧室：隐藏精彩

　　主卧室以高贵的米金点缀浓郁的红色，空间大气、奢华，彰显主人生活品质。晶莹剔透的水晶吊灯、金色麋鹿摆件等别致的软装饰品，则凸显了主人不凡的生活情趣与品位。

重现法式时光
REPRODUCING FRENCH TIME

>>> 设计理念 | DESIGN CONCEPT

The design inspiration comes from the last part *Reproducing the Time* of the book In *Search of Lost Time*. From *Reproducing the Time*, we can learn daily life details of authentic French upper class, tea party, ball, reception and other fashion social occasions; the descriptions of classic French upper class made French life famous.

Until today, French aesthetics and culture are still widely mentioned and continued. Facing heavy French civilization, the designers choose to use modern and concise lines to present a modern and light luxurious French style.

设计灵感来源于《追忆似水年华》系列的最后一部——《重现的时光》。从《重现的时光》我们可了解到原汁原味的法国上流社会的日常生活细节：茶会、舞会、招待会及其他时髦的社交场合，其中对经典法国上流社会的描写让法式生活名声大噪。

直到今天，法式的美学及文化仍广泛被提及和延续，面对厚重的法式文明，设计师选择以现代简明线条勾勒，展现一种现代轻奢法式风格。

项目名称：武汉旭辉　钰龙半岛联排别墅样板房
设计公司：赛拉维室内装饰设计（天津）有限公司
设计师：杨贵君、杨雪、许慕珠
项目地点：湖北武汉
主要材料：金属、皮质、布艺等

>>> 客厅：法式美学精髓

客厅的设计浪漫、优雅、精致，正是法式美学的精髓。灰白的几何图案地毯铺垫空间的质感，黄色皮质脚凳的点缀提亮整个空间。沙发色调、材质柔和，搭配以金属亮色的茶几，轻奢和谐，高贵典雅。客厅壁炉书柜的组合运用了法式廊柱、雕花线条，制作工艺精细考究。同样是帝黄色和金色的饰品点缀，平添了空间的味道和艺术性。

主体色
PRIMARY COLOR

点缀色
INTERSPERSED COLOR

>>> 色彩分析 | COLOR ANALYSIS

The designers choose noble and gorgeous yellow to create an unusual French residence. Yellow in high brightness and high saturation attracts eyes; from depressing gray to cool blue, the space is noble and concise, low-key and gorgeous. Champagne metal accessories enrich the colors and create a luxurious atmosphere.

设计师选择贵气华丽的黄打造非惯性法式居所。高亮度、高饱和度的黄色吸睛诱人，从沉郁的灰色到清爽的蓝色，高贵简洁中又有着低调的绚丽。增加香槟色金属饰品，丰富色彩的同时，创造奢华氛围。

>>> 地下一层：光影流转

地下一层作为主人娱乐和休闲的区域，是一个颇具气势的双层共享空间。1.5米高的水晶大吊灯，华丽璀璨。光线打在精致的家具上，突出光影在空间层次中的韵律感，增加画面的层次和质感。窗帘与家具的色调遥相呼应，两排通顶展示柜展示了主人从世界各地搜集来的名贵收藏品和书籍，带来了非凡的感官体验，凸显了独具魄力的大家风范。

>>> 餐厅：花开烂漫

客餐厅一体化的空间宽敞明亮，满足全家人就餐、聚会等功能要求。餐椅背面的布艺采用法式典型的花鸟图案，精致的餐桌饰品及花艺，展现法式的高贵优雅，每一样装饰都似乎在诉说着这个空间的魅力。

>>> 休闲空间：粉红豹的动物世界观

二层是女孩房和男孩房，中间有一个小憩的休闲空间，可以在满足亲子活动功能的同时，提供更多的畅想空间。设计师运用拟人的手法，从粉红豹的"视角"出发，描绘了一幅生动有趣的动物集会图。

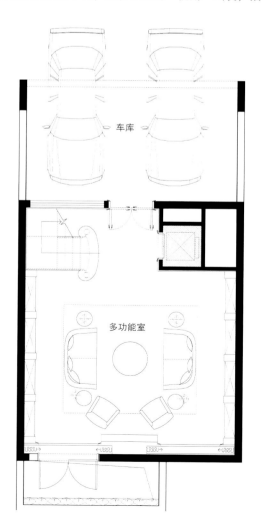
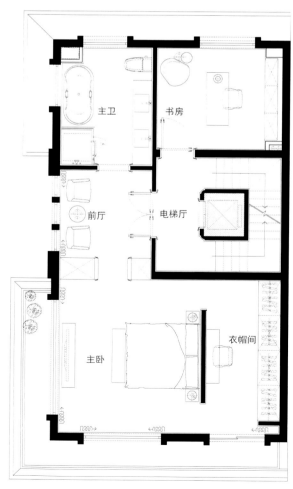

>>> 主卧室：豪华居住空间

　　三层是男女主人较为私密的空间，主卧两侧墙面均设有窗户，通风采光良好，延伸了空间视野。闲暇时坐在窗边喝杯咖啡，眺望墨水湖畔，恬静惬意。背景墙金砖堆砌的构成感和高雅床背的结合相得益彰，散发着法式与生俱来的高贵气质。床头背景墙后的衣帽间，展现了豪华舒适的居住空间。

>>> 男孩房：怀抱飞行员的梦想

男孩有一个当飞行员的梦想，床头背景的蓝天、白云、飞机充分表现了男孩对实现梦想的迫切和渴望，装饰画、装饰品、玩具都围绕这个主题展开，和谐统一。

>>> 女孩房：音乐王国

粉红色结合白色的床及床品，把女孩粉红色的公主梦变成了现实。女孩喜欢音乐，喜欢弹琴，白色的钢琴是她的最爱，演奏是每天必不可少的。床头背景墙亮眼的金色装饰好似跳跃的五线谱，紧扣空间主题。各种音乐相关的饰品摆件，打造出梦幻般的音乐王国。

夏日里的丛林世界
JUNGLE WORLD IN SUMMER

>>> 设计理念 | DESIGN CONCEPT

Living in the busy city, modern people want to pursue a soothing and peaceful environment and a home where they can rest after work, discharge all the exhaustions, forget work troubles and alleviate life pressure.

The designer uses elegant design style to deduce exquisite and fashionable contemporary French tonality. The most impressive thing is that the designer presents a strong coordination of minimalism and decorationism to us. As for home design philosophy, design is pure and comfortable and at the same time has high requirements for every furnishing. Delicate and exquisite artistic home is never lack of pursuits of artistic realm; every color presents strong personality and artistic temperament.

忙碌的都市生活，让活在当下的人们反而更加追求舒缓、祥和的环境，更加需要一个下班后能放松身心，卸下所有疲惫，忘却工作烦恼，缓解生活压力的家。

以优雅的设计风格演绎精致时尚的当代法式格调，最令人深刻的就是，设计师向我们展现了极为强烈的简约主义与装饰主义的协调性。对于家的设计哲学，设计是纯粹与舒适的，但同时又对每个装饰品都有极高的要求。精致细腻的艺术之家，总少不了对艺术境界的追求，每种色彩都展现了强烈的个性化及艺术气质。

设计公司：一然设计
设计师：杨星滨
项目地点：辽宁沈阳
项目面积：284m²
主要材料：布艺、金属、大理石等

主体色
PRIMARY COLOR

点缀色
INTERSPERSED COLOR

>>> 色彩分析 ｜ COLOR ANALYSIS

Lemon yellow furnishings full of pleasant feelings are used in the Tiffany blue and gray space; heavy and gorgeous colors make it easy to become the leader of the space and to tightly attract our eyes. Elegant and lively Tiffany blue matches with sunny and soft yellow, which is the combination of noble and romantic aura with elegant and lively temperament and the intersection of freshness and brightness with passion and flexibility. At the same time under the foil of pure color furniture and walls, it is full of texture and layering, promotes tone of the space again and again and reflects elegant breaths.

充满愉悦之感的柠檬黄色家居入住Tiffany蓝与灰调空间，浓郁而绚丽的色泽使其轻易的成为空间的主导者，牢牢吸附着我们的视线。淡雅明快的Tiffany蓝与有着阳光般轻柔触感的黄色的搭配，是高贵浪漫气场与素雅活泼气质的携手，是清新爽朗与热情灵动的交汇。同时在纯色家具与墙面的映衬下，极富质感与层次，将空间的格调一再提升，雅致气息处处展现。

>>> 客厅：灰金漫舞，演绎新时尚

淡灰色和金色搭配，打造出典雅、舒适又极具品质感的空间。金色框架配合新西兰羊毛休闲椅，让空间灵动通透。两把椅子中间的金属树墩是艺术化之后的家具，居家是我们生活的后台，更具文艺情结。壁炉处金属鹿头代替普通的装饰，设计师用现代思维重新定义经典。

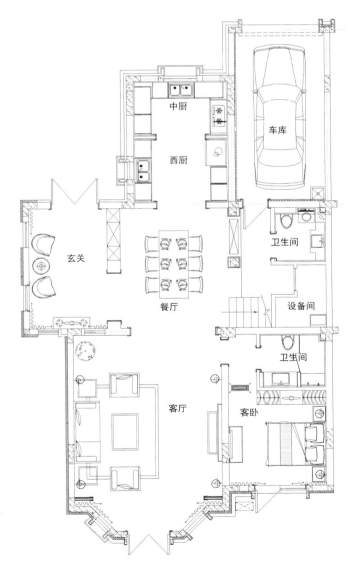

>>> 玄关：虚实应雅趣

以 Tiffany 蓝和黄色作为玄关主基调，生动的撞色，瞬间让人心情变得美美的。而巧置的灵动窗子，配上精美的的帘艺，陈设有虚掩，空间既通透又有交流。

>>> 餐厅：丛林世界里的精致浪漫

整个餐厅的设计仿若处在一个油画氛围的世界里。油画的色彩延伸到椅背上，让人身临其境，忘却钢筋混凝土堆构筑的丛林生活。溜冰状的吊灯，想起在东北小时候妈妈亲手做的灶台鱼。双色的餐椅，洁白桌面再配上橙色和绿色的叶子，自然氛围爆棚，使人心情舒适。

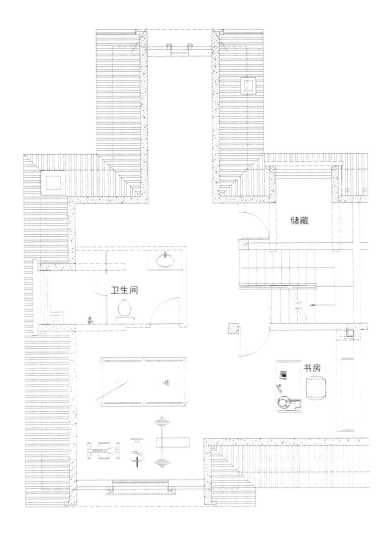

>>> 主卧：热烈与优雅兼具

　　以爱马仕橙作为空间主色调，蓝宝石色作为跳色，搭配世界顶级壁纸创造来自撒哈拉沙漠最美的鸟——蕉鹃的羽毛，艺术化的空间，奔放的色泽，皇冠造型床头，凸显高贵气质。

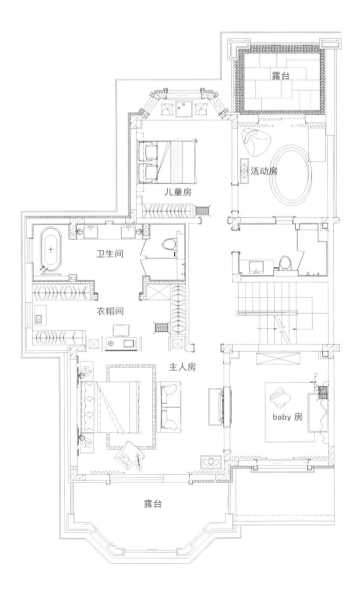

淡抹相宜 雅致相趣
CHARMING AS ALWAYS, ELEGANT AND FUNNY AS USUAL

>>> 设计理念 ｜ DESIGN CONCEPT

The quaint beauty of Oriental ink painting encounters with modern design style, getting rid of traditional symbolic piles, calm and decent. The designer abandons traditional Chinese complications and depressions to perfectly combine modern furnishings with fashionable colors so as to present an amazing beauty. It seems that we are in a piece of ink painting with thick and heavy colors among landscapes.

Implicit, meaningful, refined and graceful flower art, full, rich, fresh and lively colors, exquisite, delicate, fashionable and individualized furniture, all harmonious furnishings make the plain space full of classical and elegant flavors.

东方水墨画的古韵美与现代设计风格相遇，摆脱了传统符号化的堆砌，稳重而不失大气。摈弃传统中式的繁琐与沉闷，将现代家居与时尚色彩完美结合，呈现出一种令人惊艳到内心的美。让我们仿若处在一幅浓墨重彩的水墨画中，身处山水之间。

含蓄隽永、端庄丰华的花艺；饱满丰富、艳丽明快的色彩；以及细腻精致、时尚个性的家具。一切和谐的铺设，将朴素的空间顿时充盈了古典雅致的气息。

设计公司：一然设计
设计师：杨星滨
项目地点：辽宁沈阳
项目面积：106m²
主要材料：金属、布艺、大理石等
摄影师：盛鹏

>>> 色彩分析 | COLOR ANALYSIS

Large piece of gray in the space encounters with dark blue and noble yellow, which is as if the arrogant knight encounters with the elegant princess, falling in love with each other. Calm blue emits elegant and mysterious flavors; yellow brings people dynamic vitality as if gold rippling wheat weaves under the tranquil sky, which is mature and reaps striking affections. As a particular color, only collocating with proper colors can yellow create a perfect space.

整体空间大面积铺陈的灰调遇到藏蓝色与贵族黄，如高傲的骑士遇见优雅的公主，互生爱慕。沉静的蓝色散发着优雅且神秘的气息，黄色带给人的是活跃的生命力，犹如宁静的天空下跃动着的金色麦浪，成熟中收获着惊人的感动。但黄色作为挑剔的颜色，只有搭配合适的颜色才能打造完美的空间。

>>> 客厅：烟波浩渺现从容

以中国的贵族黄为主座，搭配一幅烟波浩渺的中国写意山水画为背景墙面，以现代的手法提纯，是一种运筹帷幄、大气磅礴。同时没有采用过多琐碎的东西，以大块大面的形式重新定义这个空间，透出一种从容、恬然之感。

>>> 餐厅：一段快乐的色曲

餐桌上的叶子和小鸟交相辉映，营造出一幅生动的自然画卷。金、橙、绿的彼此呼应，谱出一首现代东方的交响曲。同时没有用过多的中式符号去堆砌，而采用饱和色彩去丰富整个空间，更加符合当代气质。

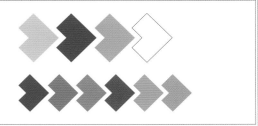

主体色 PRIMARY COLOR

点缀色 INTERSPERSED COLOR

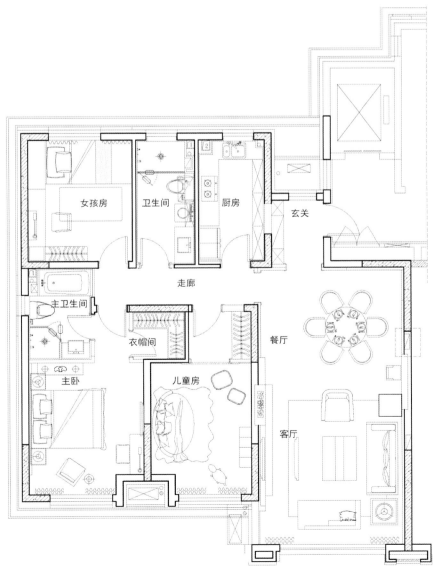

>>> 主卧：金与蓝的悦动

采用棚墙一体的设计手法，创造出温柔私密的包裹感，形成舒适而又极具安全感的空间。架子床是主人私属领地，休息的私属空间。床头柜和床头的铜，金色与蓝色，使空间典雅又大方。

>>> 儿童房：梦幻乐园

每个儿童房都应该是孩子的游乐园。树状的门、柔软的鲨鱼座椅、彩色几何的地毯、梦幻的小床、云状的灯光，这里就是孩子生活的小天堂。

>>> 学习区：小小未来星

布艺手工画、可爱的芭比娃娃、头顶会发光的红兔等，一切美好的小物件，在淡紫色背景的映衬下，都是陪伴孩子学习的小伙伴，即使他是独生娃，都不觉独孤。

华丽温馨
是你欢喜跳跃的曲子
富有南国情调
豪爽 成熟
当与白黑组合 含蓄优雅
当与蓝色组合 响亮生动
辉煌醒目间记录温暖

ORANGE

东情西韵 悦享人生
ENJOY LIFE WITH ORIENTAL AND WESTERN CHARMS

>>> 设计理念 | DESIGN CONCEPT

French new decorative style is a fashionable household style with fashionable romantic breaths and elegant tonality as the theme and has been accepted and recognized by more and more people. Nowadays, new decorative style has become the mainstream design style of domestic luxurious residential design. It presents concise lines and is injected with fashion design elements and modern decorative materials whose elegant and rigorous traits are properly exaggerated so as to reveal a slight of gorgeousness in classical charms.

As a five-story villa, the spaces are adjusted as clean as possible; irregular spaces in the house type are deleted and concluded, which enriches the shape and structure of the space and promotes life details and tastes of the owner at the same time keeps the functional location of the original architecture. The space performance is tightly close to the stylish theme of the house type, bringing elegance and rigorousness perfectly into the owner's daily life.

　　法式新装饰主义是以时尚浪漫的气息，富有典雅风味的格调为主题的时尚家居，获得越来越多人的接受和认可。在今天，新装饰主义风格已经成为了国内豪华住宅设计的主流设计风格；在呈现精简线条的同时，又为其注入了一些时尚的设计元素和现代装饰材料，并且将其优雅与严谨的特质适当地夸张，使其在古典的韵味中自然流露出一抹惊艳。

　　户型为五层别墅，户型空间在设计中尽可能地将空间调整得干净，删减、归纳户型中出现的不规则空间，在保留原建筑功能位置的同时，丰富空间的形体结构造型及附属功能区域，提升出业主生活细节及品位，空间的展示性紧紧地贴服户型风格主题，将那份优雅与严谨完美地带入到业主的家居生活中。

项目名称：北京远洋天著别墅
设计公司：北京紫香舸装饰设计有限公司
设计师：紫香舸团队
项目地点：北京
项目面积：512m²
主要材料：金箔、桃花芯木、压铜玻璃、壁纸、意大利黑金花等
摄影师：梁志刚

主体色 PRIMARY COLOR	
点缀色 INTERSPERSED COLOR	

>>> 色彩分析 | COLOR ANALYSIS

On color collocation, the designer chooses orange, white, blue and dark coffee, which perfectly presents classical and fashionable traits of the space and fully interprets the visual impacts which French new decorative style brings to viewers.

色彩上选择橘色、白色、蓝色和深咖色进行搭配设计，将空间古典又时尚的特质表现得恰当好处，充分诠释法式新装饰主义风格带给观者的视觉冲击力。

>>> 主题沙龙区：魅力沙龙

沙龙区自在随意，衬以姜根色墙身，以鲜明的橘色长条沙发为一室焦点，一鲜艳一低调的色彩搭配形成和谐对比，营造出充满活力的空间气氛。以镜面装饰增加空间的尺度感。沙龙区后即为酒窖，饰架上陈列着来自世界各地的美酒佳酿，方便沙龙聚会中的宾主把酒言欢。

>>> 餐厅：超现实与复古风味的交错

餐厅上方的圆顶设计，凸显主宰四方之感。蓝橘颜色交错搭配更具有视觉冲撞性，一个鲜明，一个沉静，表现出现代与传统、古与今的交汇，碰撞出兼具超现实与复古风味的视觉感受。金色的贝壳靠背亮眼醒目，蕴含饱满闪耀的光芒，散发着恒久的高贵色泽，赋予空间一种新的生命力。

>>> 起居室：糅合东方情怀与现代风尚

　　进入起居室，色彩斑斓的祥龙升天屏风随即映入眼帘，配合蓝色几架，于东方情怀中流露几分跳脱的现代风尚。空间中"一半"是静谧的蓝，"一半"是亮丽的橘，家具与饰品的完美结合，冷暖色调的依量对比，蕴涵着浪漫、华贵的气息。

>>> 卧室：缤纷多彩的休憩空间

卧室贯彻温馨雅致的风格，大多以白色打底，或是秀丽清新的湖蓝、或是沉静安详的灰蓝，或是瑰丽温暖的亮橘，成就高雅时尚的休憩空间。儿童房则显得活泼生机得多，以经典英伦元素为空间要点，打造出充满激情与动感的乐园。

英伦风情 品位经典
BRITISH STYLE, CLASSIC TASTE

>>> 设计理念 | DESIGN CONCEPT

Starting from the overall construction of the location of the villa, the designers deepen the function layout of the original show flat to greatly promote the utilization of the space. In the first floor, they integrate Chinese and Western kitchens into a whole to make it an open kitchen so as to make the entire floor transparent and magnificent with visual senses. The original layouts of the second and third floors are too trivial and the function partitions are not clear. So the designers boldly break the traditional layout of the villa that children's room should be in the second floor and master bedroom the third floor. They redesign the original room distribution to make children's room in the third floor and master bedroom the second floor.

Combing with the architecture form of the British style villa, the interior design gives priority to British style to create a private space where you can quietly and elegantly enjoy the British afternoon tea.

设计师从别墅所在区域整体建设出发，深化原样板间功能布局，令空间的使用率大大提升。首层，将原有的中西厨合二为一，变为开放式厨房，使得整个首层空间更加规整通透，视觉感觉更加大气。二层和三层的原有布局太过琐碎，而且功能分区不明晰，因此设计师大胆打破别墅传统的二层儿童房、三层主卧的格局，反其道而行之，将原有房间重新分布设计，改为三层儿童空间、二层主卧套房。

结合别墅英伦风格的建筑形态，室内设计上主打英式风情，营造出了一个可以安静、优雅品尝英式下午茶的私家空间。

项目名称：北大资源·阅府155平方米别墅样板房
设计公司：赛拉维室内装饰设计（天津）有限公司
设 计 师：王少青、杨贵君、张为彬
项目地点：天津
项目面积：155m²
主要材料：木质、金箔、布艺等

主体色 PRIMARY COLOR	
点缀色 INTERSPERSED COLOR	

>>> 色彩分析 ｜ COLOR ANALYSIS

The soft decoration colors match orange and blue with yellow; artistic color collocations endow the space with dynamic rhythm and aesthetics. Coffee, dark blue and wine red make people relaxed and comfortable and calm people's inner heart down to savor every detail of this house.

在软装色调上，主要以橙色、蓝色、搭配中黄，富有艺术色彩的配色处理赋予空间动态的韵律和美感。而咖啡色、藏青、酒红等沉稳的色调给人放松舒适的感觉，让人内心沉静下来，细细品味这所房子的每一个细节。

>>> 会客厅：浓郁英伦风范

　　一层打造英式会客厅，顶面使用 BURBERRY 标志性的菱形格子元素，配以镜面材料，使空间更加高挑、明亮。拼花造型的大理石地面，与顶面相呼应。主沙发背景墙面喷画为英国剑桥大学内剑湖的一处景观，对面整墙为深色木作书柜，充满了浓郁的英伦氛围。

>>> 餐厅：与君同醉

　　餐厅与客厅布局分明的同时保持连贯性。六人方形餐桌，搭配米黄色皮革座椅，轻奢大气中增添温馨。方形天花装饰以铜质吊灯，点点光芒洒在漆面餐桌及金色墙饰上，营造瑰丽氛围。陈列架上的各式酒杯及酒瓶，流露出户主对酒文化的非凡品位。

>>> 三层空间：专属儿童乐园

　　三层打造出一个完全属于孩子的空间。利用层高的优势，进行搭建和改造，规划出睡眠区、游戏区、学习区、收纳区、盥洗室等多重功能区。在空间搭建出二层船型迷你阁楼，营造出浓郁的航海风，给了孩子一个探险的奇妙空间。橙色和蓝色相互碰撞，表现出强烈的英伦色彩。

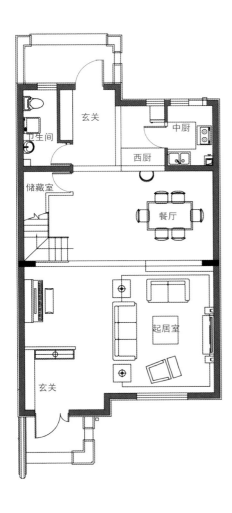

>>> 主卧与书房：奢华有道

二层空间改造最大，将房间进行拆改，化零为整。以主卧为中心，打造出主人房套间，书房与卧室一体化、通透化的设计，使得空间更加开敞，更适合展示和居住。衣帽间面积更大，收纳空间更丰富，卫生间功能也更齐全。

空间定制地板沿用 BURBERRY 标志元素，镶嵌玫瑰金边，低调奢华。大面积运用橙蓝色调冷暖对比，给人留下深刻的印象。书房装饰画选择了具有狩猎主题和英伦服装的作品，与书房氛围更为贴合。

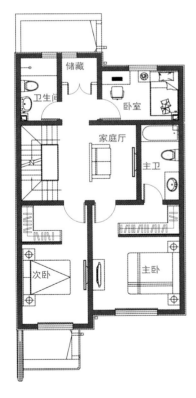

摩登时尚 自在之家
MODERN FASHION, COMFORTABLE HOME

>>> 设计理念 | DESIGN CONCEPT

Shanghai Yanlord Century Park is a high-end project costly created by Yanlord, located in the core area of Pudong Huamu, only 800 meters straight to the "Century Park" within Shanghai inner ring roads. This project uses Hermes style throughout the entire living space to make it reveal graceful and luxurious feeling in low-key charms. As a popular color scheme, Hermes orange is appreciated by many designers and owners. The designer pursues practicability and comfort of the entire space at the same time manifests modern senses of brands and perfectly interprets the unique life attitude and concept of Hermes, namely freely enjoying fun brought by life in low-key luxury.

上海仁恒公园世纪是仁恒置地斥巨资倾力打造的高端作品，项目位于浦东花木核心区域，仅800米直线距离上海内环线内"世纪公园"。本案采用了爱马仕风格贯穿整个家居空间，使其在低调的韵味中透露着雍容和奢华之感。爱马仕橙作为流行的配色方案得到许多设计师和业主的青睐，设计师在追求整体空间的实用性与舒适感的同时，也彰显品牌的现代感，完美地诠释了爱马仕独有的生活态度和理念，即在低调奢华中自在享受生活带来的乐趣。

项目名称：仁恒公园世纪
设计公司：奥妙陈设
主创设计师：朱芷谊
项目地点：上海
项目面积：200m²

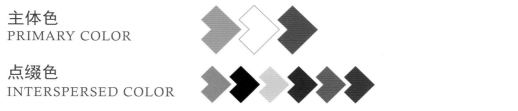

>>> 色彩分析 | COLOR ANALYSIS

This project uses modern fashionable Hermes orange throughout the space, adding a warm feeling for the owner even in autumn and winter. No matter as the main role of the living room or the supporting color and interspersed color in dining room and bedrooms, it meets people's aesthetic needs of modern life; its gorgeous and bright, low-key and luxurious, classic and fashionable labels become eternal luxurious tastes. Even it collocates with other background colors or interspersed colors, such as elegant gray, sapphire blue and jazz white, it retains its luxury and exquisiteness.

本案使用潮流时尚的爱马仕橙色贯穿整个空间，即使在秋冬也能为业主增添一丝温暖感受。无论作为客厅的主角色，还是餐厅卧室的配角色和点缀色，它都符合人们对现代生活的审美需求，其艳丽明亮、低调奢华、经典时尚的标签，成为永恒不变的奢侈品味。搭配其他背景色或者点缀色，如雅灰、宝蓝、爵士白等，也不失奢华与精致。

>>> 客厅：欢快贵气 不失时尚

客厅采用爱马仕橙的家具和饰品搭配，不同材质和颜色的对碰，让空间呈现出一种只有上层人士才会拥有的生活方式。浓烈爱马仕风格挂画汇聚客厅的焦点，让空间变得摩登贵气又欢快十足。长条沙发和单椅以米白色为主，同时使用橙色作为点缀，再搭配茶几上橙色的插花，使得空间因橙色串联起来变得有序而和谐。

>>> 餐厅：雅灰之上 橙色相配

餐厅墙身使用浅雅的银桦灰地砖作为基础用色，而用一面是亮黑一面是显眼明丽的爱马仕橙餐椅作为主角色彩，以橙色餐具和餐巾作为餐桌上的点缀，不仅和餐椅相呼应，也让爱马仕橙色在灰色调之下更加凸显出时尚的质感。

>>> 主卧：异域风情 沉睡天明

　　主人房以白色的天花为背景色，深橙色的床头背景墙平分了墙体一半的色彩，橙白相对，形成跳跃的视觉感受。床位挂画也巧妙选用宝蓝色带来的跳跃感，成为空间里无法忽略的一抹异域色彩，爱马仕橙和宝石蓝形成强烈对比，在丰富色彩层次之余，分外地迷人。

>>> 男孩房：活力四射 张扬青春

以白色打底，既是环绕空间的背景色，也是占据视线的主角色，如此纯净的卧房恰到好处地描绘出孩子天真无邪的心性。但床头挂画和床上的棒球又透露出男孩阳光活力的青春状态，深蓝的床毯点缀得醒目明朗，既是色彩上的补充，又诠释出男孩活泼洋溢的性格。

爱马仕时尚家
HERMES FASHION HOME

>>> 设计理念 | DESIGN CONCEPT

As is known to us all, Hermes is a famous French fashion and luxury brand, has upheld the extraordinarily excellent and extremely gorgeous design concepts and creates extremely elegant traditional model. With the profound understanding of "international" living culture, the designers absorb artistic temperament of international fashion brand and infuse exquisite modern materials, leather chairs, modern furnishings and metal texture to make a harmonious coexistence of exquisiteness and fashion. They also combine Oriental aesthetics and integrate Chinese classical elements such as carved tables and chairs, tea sets, ancient books and lamps to convey cultural flavors of Chinese style.

The designers use a Chinese and Western fashion feast to deduce a legendary Chinese story to create a magnificent and elegant "Hermes fashion home" which perfectly presents an exquisite and delicate living space.

众所周知，爱马仕（Hermes）是法国著名时装及奢侈品的品牌，一直秉承着超凡卓越、极至绚烂的设计理念，造就优雅之极的传统典范。带着对"国际范儿"居住文化的深刻理解，设计师汲取国际时尚品牌的艺术气质，运用精细的现代材料，皮革座椅、现代风格饰品与金属质感融合，精致和时尚和谐共生。结合东方美学，融汇中国古典元素的雕花桌椅、茶具、古书、台灯等，传达出中式风情的文化韵味。

设计师以一场中西合璧的时尚盛宴演绎出传奇的中国故事，打造出一个大气优雅的"爱马仕时尚家"，完美地呈现出精致有情调的居家空间。

项目名称：金地江南逸98平方米样板房
设计公司：杭州易和室内设计有限公司
设计师：李扬、祝竞如、马辉
项目地点：浙江金华
项目面积：98m²
主要材料：不锈钢、墙纸、皮革等
摄影师：阿光

主体色
PRIMARY COLOR

点缀色
INTERSPERSED COLOR

>>> 色彩分析 | COLOR ANALYSIS

There is no doubt that orange is the brightest color of this project; it brings the inherent energy and fashion which occupy the viewers' visions. It collides furiously with blue green so as to become more dynamic. Gray and champagne metal color intersect in it, promoting visual charms of the space.

　　橙色无疑是这个案例中最亮眼的颜色，它带着与生俱来的活力与时尚占据着观者的视野。与蓝绿激烈碰撞，显得鲜艳而具有活力。灰色与香槟金属色穿插其中，提升空间视觉魅力。

>>> 客厅：华丽旅程

　　这是一位名流绅士的华丽旅程：牵一匹骏马穿越奇花异卉的丛林，迎着阳光活力的橙色之光，追寻"国际范儿"的生活本质。东方巧匠们的手工艺品精妙绝伦，一抹亮丽的中国红绝美祥和。现代与古典、科技与手工艺、西方与东方的完美结合，最终圆满了这个新东方的时尚情怀。

>>> 餐厅：古典与现代共生

　　隔着花，望着画，坐在古典的中式雕花桌椅上，享用浪漫的晚餐。古典的韵味配上点现代的元素，美味里也增添了不少艺术的气息。

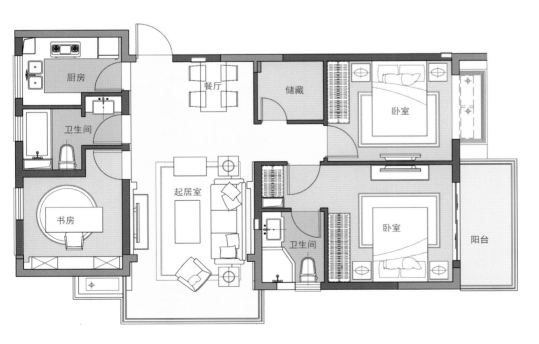

>>> 主卧：骏马归野

若说这屋里最生感情、最寄予情怀的空间就是卧室了。一抹灰色，庄重典雅，抚平浮躁的心灵；几缕爱马仕橙色，个性可爱，品味时尚的潮流。月明之时，天空皎洁，骏马归野，为生命而歌，正合时下禅意人生。

>>> 小孩房：丛林历险

对于立志成为少年特种兵的孩子来说，没有比迷彩更合心意的颜色了。惊心动魄的丛林历险，勇于挑战的故事，无疑是一个快乐的陪伴。

墨绿轻奏 红橙浅唱
DARK GREEN PLAYS LIGHTLY, RED ORANGE SINGS GENTLY

>>> 设计理念 | DESIGN CONCEPT

This project uses decorative details with Oriental flavors, complemented by modern design technique, manifesting exquisite details and making luxurious charms encounter unexpectedly with passion. Furnishings with Oriental flavors narrate the living fun of traditional Chinese tranquil and leisure literati; Zen flower arrangement, Chinese furnishings and dark green bedside background wall outline Chinese elegant and graceful spirits in a casual way. The tensions of modern bold color collocations make the dark green become more striking and the red orange more gorgeous. The application of metal materials and white and dark color irregular grain carpet make people jump out; Chinese and modern clear lines decrease the deliberated design sense in conciseness and increase life sense, which is the hidden design.

本案以东方风韵为装饰细节，配衬现代的设计手法，细节间的缠绵尽显精致，让奢华魅力与激情不期而遇。具有东方气息的摆件将传统中国恬静闲适的士大夫居住趣味娓娓道来，禅意插花、中式摆件、墨绿床头背景墙，于细节处不经意勾勒出中式文雅隽秀的精魂。现代大胆配色之间的张力，让墨绿色绿得更加醒目，使红橙色显得更加亮丽。金属材质的运用，白色和深色不规则纹理的地毯让人跳跃，中式和现代风格的线条分明，于简约中让刻意的设计感减少，生活感增多，这便是深藏不露的设计。

项目名称：公园里样板房贰
设计公司：广州杜文彪装饰设计有限公司
设计师：杜文彪
项目地点：北京
项目面积：150m²
主要材料：金属、实木复合地板、天然石材等
摄影师：Bill

主体色
PRIMARY COLOR

点缀色
INTERSPERSED COLOR

>>> 色彩分析 | COLOR ANALYSIS

Spaces of this project use colors to create a basic tone of high lightness and low stability; creamy white ceiling as the main role of the space makes the upper part cooler and extends the space upward. Spaces with green and orange as main roles collocate with dark or light background colors, such as dark coffee living room, dining room and bedrooms, black and white daughter's room and beige second room, creating a more composed and steadier living atmosphere to make owners feel comfortable and leisure, free and cozy.

　　本案空间通过色彩营造出上轻下稳的基调，以米白色天花为主角色让空间上半部分显得清爽，让空间得以向上延伸。以绿色和橙色为主角色的空间搭配或深或浅的背景色，如深咖色的客餐厅和卧室、黑白相间的女儿房、米黄色的次卧等，使之营造出更加沉稳而踏实的居住氛围，让业主居于此舒心解意，自在安心。

>>> 客餐厅：高雅摩登

客厅地毯黑白规则曲线，宛如婀娜多姿的女子，风姿绰约。红橙与墨绿的邂逅，打破平淡如水之境，色彩点燃的激情一蹴而就。空间的奢华邂逅历史的沉淀，高雅又不失摩登。餐厅味蕾狂欢之夜，舌尖碰撞，红橙墨绿，光影交替，犹如星空散落其中，流光溢彩。虽无丝竹管弦盛乐，一觞一咏，亦足以畅叙幽情。

>>> 主卧：清幽古香

　　古色的木质墙体为背景色，在灯光的照射下，似乎散发出清幽古香。浅橙色的主角色床品以大面积占领视觉中心，墨绿色的床头背景墙与床尾挂画摆件相呼应，绿色的抱枕在橙色中加以点缀，灯光洒落，柔和中带着温暖，卸去装束，释放本性，千彩千姿如梦般沉醉。

>>> 次卧：紫气东来

　　竖条斑马纹，黑白交错，原始的野性中略带着一点温柔可人的味道。紫色的主角色让穿插在空间中跳跃的黑白色得以抑制，减少了冷色调的清冷，加之点缀在卧室中的红色和橙色，使暖色的温度和热情得到释放，让居住者在色彩搭配中体会空间平衡的妙处。

>>> 女儿房：黑白魅惑

女儿房黑白携手，跳跃有力度，简约内敛的色调彰显着优雅知性气质，灯光交织，令人心醉沉迷的除了玛丽莲梦露经典白裙般的魅惑，还有奥黛丽赫本小黑裙般的优雅。浓妆艳抹之下，笑容的背后，也只是一个眺望世界的小女孩。

橙色魅力 至美之家
CHARMING ORANGE, BEAUTIFUL HOME

>>> 设计理念 | DESIGN CONCEPT

According to the location and show flat characteristics of this project, the design team focuses targeted clients on "Mr. and Mrs. Smith" in Tianjin when designing this show flat. Reasonable space planning highlights qualities and characteristics of this project, bringing perfect living experience.

The targeted family structure of this show flat is a family of three, with living room and master bedroom in the south, daughter's room, multi-functional room, dining room, kitchen and living balcony in the north. The multi-functional room is designed in open pattern and links with the living room, forming a north-south transparent house type. The design style adopts Hong Kong light luxurious style with favorite Hermes as the theme; rose metal, soft coverage and leather present a luxurious quality; mirror design elements are used to expand visual feelings of the space.

针对该项目所在区域及样板间特点，设计师团队在设计样板房时将客户群定位在天津的"史密斯夫妇"，合理地规划空间，凸显项目的品质和特点，带来绝佳的居住体验。

该户型客群家庭结构为三口之家，南向空间为客厅、主卧室，北向房间设置为女孩房、多功能房、餐厅、厨房、生活阳台。多功能房设计为开敞式设计，与客厅从空间上形成贯穿，从而实现南北通透户型。设计风格上，采用港式轻奢风，以钟情爱马仕为主题，用玫瑰金属、软包、皮革等材质体现奢华品质，并运用镜面设计元素来扩大空间视觉感受。

项目名称：天房·观锦128·三居样板房
设计公司：赛拉维室内装饰设计（天津）有限公司
设计师：王少青、杨贵君、张为彬
项目地点：天津
项目面积：128m²
主要材料：金属、软包、皮革等

>>> 画作：阿房宫图

画作节取清代画家袁江《阿房宫图》部分。设计师将富丽美观的山水画作移至家中，强调气势恢弘的视觉效果，使一组组已经逝去的带有神秘色彩的建筑得以再现。建筑物用大青绿敷色，浓艳厚重，鲜艳夺目。山石树木则用水墨，略施淡彩，主次分明。雄伟壮阔的山色，富丽堂皇的楼阁，很好地融为一体，既精细入微，又气势磅礴。

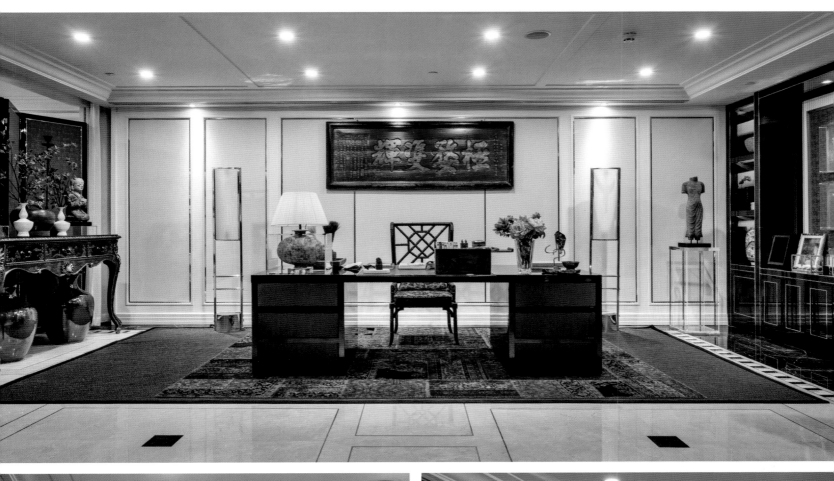

>>> 主卧室：凤凰于飞 翙翙其羽

当黑色与金色相遇，极致的华丽之光难以掩藏。黑金的搭配既反差又融合，散发出让人无法忽视的魅力属性。时髦复古黑金背景给人深沉肃穆之感，雕琢大气的金色凤凰图腾作为点睛之笔夺人目光。

>>> 次卧室：繁华褪尽的温暖

卧室极具方正对称美，质朴大气。设计师特意选取深咖色的床头柜与电视柜相呼应，体现空间设计的整体性。龙鳞纹床品既是中式风格的典型体现，也是居者内在气度的升腾。

锦绣满堂
SPLENDIDNESS

>>> 设计理念 | DESIGN CONCEPT

She is romantic, noble and passionate and sends out charming amorous feelings everywhere. She comes from a romantic country, France. In the afternoon, she sits quietly in the coffee shop near the Seine to taste life; at dusk, she silently gazes at the scenery of the Eiffel Tower; at night, she lightly tastes Bordeaux wine, maybe she will see lavenders in Provence and visits the gorgeous Versailles palace. On holiday, she will enjoy the sunshine and beach in Marseilles. She is the most dazzling flower in the rose garden. In the romantic summer, why not be in love with her in France?

The designer pursues the whole poetic artistic conceptions, strives to have a deep influence on people from the perspective of temperament and creates a noble and free, romantic and passionate French neo-classical style on the whole through partial outlines of French elements.

她，浪漫、高贵、热情，处处散发着迷人的风情。她来自一个浪漫的国度——法国。午后，安静地坐在塞纳河边的咖啡店品味人生；傍晚，默默地凝视着巴黎铁塔的风景；午夜，浅浅地品尝波尔多的美酒，或许还会去看看普罗旺斯的薰衣草，凡尔赛宫殿的华丽。假日，享受马赛的阳光沙滩……她，是玫瑰园里最耀眼的那一朵花蕾。浪漫夏日，和她谈一场法国恋爱吧！

设计师追求整体的诗意意境，力求在气质上给人深度的感染，通过局部法式元素的勾勒，从整体上营造出高贵自由、浪漫热情的法式新古典风格。

项目名称：绿城·青岛玫瑰园法合别墅
设计公司：浙江绿城家居发展有限公司
软装设计师：熊萍萍
项目地点：山东青岛
项目面积：573m²
主要材料：金属、玻璃、金箔、大理石、布艺等
摄影师：三像摄影

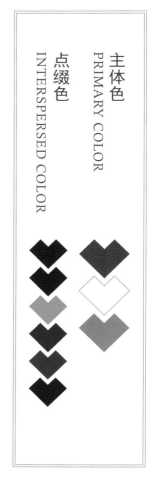

主体色 PRIMARY COLOR

点缀色 INTERSPERSED COLOR

>>> 色彩分析 | COLOR ANALYSIS

Red with passion and vitality links every space, overflowing fresh vitalities. Bright and elegant white, low-key and noble gold, blue as quiet as lake water and burgundy wine red whose color and name both make people enchanted together to compose a love song. Whether it is delicate and charming flower of the bedside lamp patterns or a shaking leisure chair by the window, in every corner, you can taste the elegance, notability and romance of French style.

带着热情与活力的红串联起了各个空间，满溢着鲜活的生命力。皎洁优雅的白、低调贵气的金、如湖水般沉静的蓝、颜色与名字一般让人心醉的勃艮第酒红，共同谱写一曲关于爱的赞歌。不论是床头台灯图案中娇艳的花朵，抑或是窗前的一把微微晃动的休闲椅，在任何一个角落，都能体会到法式风格的优雅、高贵和浪漫。

>>> 客餐厅：灿若玫瑰

打造一个激情而富有活力的客餐厅，选择饱满而浓郁的勃艮第酒红作为主题色，带着些许的性感与暖暖的优雅。金色家具的融入提升整个空间的奢华度。空间中自由而灿烂的盛放花束，或皎洁如玉，或热情如火，都绽放出永恒的生机美感。

>>> 主卧：幽若蓝星

大开窗的设计，让阳光能更多地涌入。辅之以蓝色壁纸和床尾沙发，仿佛进入无边静谧的大海。俏皮的橘红，素雅的高级灰，以点缀的形式加入卧室空间，勾勒出浪漫的动感。

>>> 男孩房：秀若芝兰

在男孩房大面积地使用蓝色，使空间更显清新素雅，渲染出宁静又不失高贵感的空间感受。米色布艺饰品和白色线条的室内装饰，突显出室内简洁优雅的氛围。

>>> **老人房：淡若雏菊**

　　黛色和米色的运用让整个心灵都安静平和，而棕色家具的加持更是使得这种感觉愈发浓重。因而此时需要做的便是将这种涤荡心灵的效果推向极致。与阳光同色的窗帘，如同枝桠间散落的斑驳阳光，柔和的色泽中带着温暖人心的触觉。

>>> 女孩房：清若水仙

淡淡的冰淇淋粉、静谧蓝、清爽绿，为整个空间注入甜美梦幻的新活力，仿佛空气里都充满冰淇淋般冰爽的清新味道。凝神细看，点点珍珠泛着润泽的光芒，不经意展现即将成长为窈窕淑女的优雅和柔美。

绽放的英伦玫瑰
BLOOMING BRITISH ROSE

>>> 设计理念 | DESIGN CONCEPT

The designer integrates neo-classical decorative style with classic British elements to perfectly present the unique layout structure of the project and highlights decent and warm noble tonality and comfort and romance of life. Perfect dots and lines and excelsior detail treatments bring endless comfortable tactility. The combinations of a variety of lamps and lanterns including droplight, wall lamp and spotlight are indispensable, creating different effects of light and shadow. The choice of furniture is also vital; their colors are mellow and warm; the collocations of different carpets and furniture greatly promote taste of the space. Vivid oil paintings, multilateral mirrors and multi-folding curtains are dotted in different areas to create atmosphere for the space. The decorative elements are harmonious and unified and integrated into a whole; staying here, it seems to gaze at the shadow of a British gentleman, what you see are elegance, dignity and leisure.

以新古典装饰风格融入经典英伦元素，独特的轮廓将结构本身发挥到极致，着重强调出大气温馨的贵族情调，还有生活的惬意和浪漫。完美的典线，精益求精的细节处理，带来不尽的舒适触感。装饰中大吊灯和壁灯以及射灯的多种灯具组合，不可或缺，营造出不同光影效果。家具的选择也至关重要，色调醇厚而温暖，不同地毯与家具的搭配，大大提升了空间的品位。生动的油画、多边境面和多重褶皱的窗帘，点缀不同功能区域的环境氛围，装饰元素和谐统一、融为一体，驻足期间，仿佛注视着一个英伦绅士的背影，看到的是优雅尊贵和从容不迫。

项目名称：嘉兴香格里拉
设计公司：王庭装饰
项目地点：浙江嘉兴
主要材料：大理石、木质板、布艺等
摄影师：逆风笑

| 主体色 PRIMARY COLOR |
| 点缀色 INTERSPERSED COLOR |

>>> 色彩分析 | COLOR ANALYSIS

Classic charm leads people to endless aftertastes. When wood fragrant red brown encounters with luxurious European style, the noble momentum is beyond words. Red brown is used throughout every space with its consistent calm and low-key traits; crooning or whispering, it brings us to review classical European romance in modern metropolis and deduces exquisiteness and aesthetics to the extreme.

古典的韵味让人回味无穷，当木质芬芳的红棕色遇到奢华的欧式，尊贵气势无以言表。红棕色以其一贯沉稳低调的特质，贯穿于各个空间，或低吟或浅唱。带我们在现代都市里重温古典的欧式浪漫，将精与美演绎到极致。

>>> 客厅：雕刻的史诗

奢华似乎一直未曾摆脱金色的烙印，然而设计师却另辟蹊径，以木香冉冉的红棕色为基底，搭配透亮的白，优雅的灰与蓝，或深或浅、或浓或淡，将欧式的精华，在流动的曲线间尽情演绎。

>>> 餐厅：简约而不简单

华丽的雕刻、庄重的格调、以及对艺术设计无穷无尽的幻想，让餐厅空间沐浴在大自然的馈赠中。在这里，无需复杂的色彩勾勒，简单的红白灰，你可以体会到真正的舒适与开阔。

>>> 厨房：味蕾挑逗

食物的美好永远无法用语言诉说，人们对它的喜爱不仅仅止于温饱，而一间适合的厨房就是制造这一切美好的源泉。全木质的顶面，全套定制的实木橱柜，还有灵活多用的中岛吧台，似乎在一转一回头间都可以闻到美食的芳香。

>>> 主卧：奢华篇章

以红棕色的木质为背景色，墨绿点缀其中，从而让沉稳安静的卧室空间增加悦动活力和浪漫气息。而白色和黑色的加入，在消解了空间厚重感的同时，还提升了时尚炫酷的奢华魅力。

玛莎拉红 高贵灵韵
MARSALA RED WITH NOBLE AURA

>>> 设计理念 | DESIGN CONCEPT

Greentown Changzhi Island Jade Garden uses the soft decoration style which combines "western technique" with "modern technique", retains the essences of French style at the same time simplifies the original excelsior characteristic elements with modern colors and design methods and presents a noble, composed, decent and restrained tone. It comes here from France with fragrance of time, inherits elegant charms of French style, transplants French romance, captures life texture by soft decoration symbols and creates a perfect living space which conforms to the identities and living tastes for elites who live in this seaside city.

Jade Garden Show Flat faces the sea and is closest to nature; no matter which corner you are in, you can appreciate the perfect scenery with green lake and blue sky; the flying sea gulls, the jumping waves and the sparkling sea can bring you incomparable holiday mood.

绿城长峙岛翡翠苑，采用了"西式技法"与"现代技法"相结合的软装风格，保留法式精髓的同时将原本精雕细琢的特点元素简约化，配以现代的色彩及设计手法，呈现出高贵、沉稳、大气、内敛的格调。它从法国跨越岁月的醇香来到这里，传承法式风格优雅的神韵，移植法兰西式的浪漫，用软装符号捕获生活的质感，为这个海边城市的巅峰人群打造真正符合身份、居住品味的第一居所。

翡翠苑样板房面朝大海，与大自然无限贴近，无论你身处在哪个角落，都可欣赏到碧海蓝天的绝佳景致，飞翔的海鸥，跳跃不定的浪涛，波光粼粼的海面，带给你无与伦比的度假心情。

项目名称：绿城·舟山长峙岛翡翠苑
设计公司：浙江绿城家居发展有限公司
设计师：熊萍萍
项目地点：浙江舟山
项目面积：600m²
主要材料：水晶、金属、大理石、布艺、胡桃木等
摄影师：三像摄

| 主体色 PRIMARY COLOR |
| 点缀色 INTERSPERSED COLOR |

>>> 色彩分析 | COLOR ANALYSIS

The target customer of this project is the higher lever group. The designer uses modern technique to present French elegance; as an interspersed color of the entire show flat, Marsala Red is used throughout the main spaces. No matter large chunks or small pieces of furnishings, dark colors always can give people a luxurious and noble temperament, elegant, mature and mellow.

本案设计目标定位为较高层次客户群，用现代的手法去表现法式的优雅，玛莎拉红作为整个样板房的点缀色，贯穿样板房的主要空间。不管是大块运用还是小件配饰，暗调的色彩总能给人一种豪华的贵族气质，优雅成熟，细品醇香。

>>> 客厅：点红新装

每一种颜色都有它的特殊意义，将热情似火、奔放向上的红色，搭配简洁优雅、纯真脱俗的白色都会有令人惊喜的效果。即使只是一把描金扶手沙发、一只抱枕、一盏台灯，也可带来浓郁的神秘华贵气质，隐约的红色点缀有着难以描绘的吸引力。

>>> 起居室和餐厅：中法融汇

设计师运用经典壁炉、法式雕花、吊灯等素材，佐以梅花图案地毯、古董花瓶、中式茶具，以白色和米色为背景色，玛莎拉红点缀其中修饰，大大增加了空间的高贵感和优雅韵味。

>>> 书房：岁月芬芳

铺满全室的玛莎拉红色地板，低调奢华的美感幻化而出。与墨绿色皮质沙发碰撞，增加空间的视觉冲击力。玛莎拉红亦覆盖了整个灯饰表面，加温了灯光的温度，为原本方正的书房加入了一种甜腻的美感。

>>> 影音室：灯色朦胧

玛莎拉红地毯有种铺天盖地的奢华感，醇厚、温暖的色调似乎从脚底蔓延至全身。变幻莫测的灯光下，有种朦胧的美感。美丽的天花镜面装饰，起到延伸空间的作用，给人多维度的视觉效果。

品享生活
ENJOYING LIFE

>>> 设计理念 | DESIGN CONCEPT

This project echoes with the Chinese cultural theme position of the building, complements by Chinese elements with modern technique and chooses "palace" as the story clue to organize decorative elements. Black and white as the main tones intersperse palace red, creating a mysterious and luxurious palace atmosphere and trying to reflect elegant and luxurious side of Chinese culture. Oriental design languages in the modern space manifest Chinese charms in details and oriental implication and flexibility in abstraction; the general part is paved with Chinese palace red while the small part reveals Oriental elegant fun. No matter traditional Oriental or modern elements, the designer puts them in the space to create a comfortable living atmosphere; elegant, concise or modern, all are good choices for life.

本案呼应楼盘中式文化主题定位，以现代手法辅以中华元素，选择"宫廷"为故事线索组织装饰元素。灰黑色为主色调点缀宫廷红色，营造神秘华贵的宫廷氛围，尝试表现中华文化优雅华贵的一面。现代空间中的东方设计语言，于细节中见中式的韵致，于抽象中见东方的含蓄灵动，大处落笔铺就中国宫廷红，细处收笔洞见东方雅趣。不论是传统东方还是现代风格的元素，设计师将它们放置空间中只为打造更加贴合人心的居住氛围，可以清雅，抑或简约摩登，都是上佳的居住选择。

项目名称：方圆湛江云山诗意 C1户型示范单位
设计公司：陈列宝室内建筑师（深圳）有限公司
设计师：陈列宝、李程
项目地点：广东湛江
项目面积：120m²

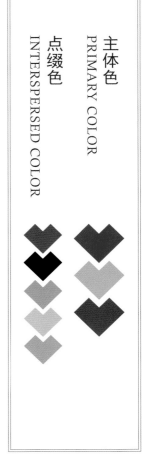

主体色 PRIMARY COLOR
点缀色 INTERSPERSED COLOR

>>> 色彩分析 | COLOR ANALYSIS

As main role of the space, red not only connects every space, but also reflects luxury of the palace. Gray black marble tiles and creamy white wall background foil charms of red; at the same time, applications of modern metal materials and mirror decoration increase the space brightness; dark coffee curtain skillfully foils the spaces and balances the colors.

  红色作为本案空间的主角色不仅串联着每个空间，同时也表现出宫廷中的华贵。灰黑色的大理石地砖和米白色的墙体作为背景色，更加衬托出红色的魅力，同时现代金属材质和镜面装饰的运用提升了空间亮度，深咖色的窗帘也巧妙起到衬托空间、均衡色彩的作用。

>>> 客厅：寒梅飘香宫廷晚

进入客厅视觉便被背景墙上大胆醒目的枯枝梅花造景所吸引，在深灰色的大理石地砖和墙面的环绕下，红色地毯和背景墙显得越发明目。米白色的沙发不争不抢点缀其中，在对比中多了一份素净，两盏高挑的落地灯造型上拟古宫廷夜灯，熠熠生辉，挑灯向晚。

>>> 餐厅：邂逅红与黑

红与黑的碰撞在餐厅上演，红色的热情和黑色的神秘之间形成的张力，让暖色更张扬，冷色更理智。布艺纹花的餐椅和红色挂画不经意间相呼应，黑色靠枕上的花优雅冷冽，餐桌上陶艺花瓶中含苞待放的梅花，还未盛放似乎已散发出阵阵幽香。

>>> 书房：红袖添香夜读书

书房是清净圣洁之地，桌上的笔墨纸砚无意中散发出缕缕清香。红色书架以夺目之势成为书房的主角色，隔层中的摆件在黄昏色的灯光中活灵活现，书架中央清雅的刺绣在一片宫红中脱颖而出，浓处着色淡雅得刚刚好。

>>> 卧室：月暗灯微欲曙天

轻踏脚步进入卧室，米白主角色的床品轻柔适眠，地毯和床柱用红色点缀并将各空间串联起来，最值得回味的是床头挂画，正经端坐的宫廷女子传递出古代女子端庄典雅之美。床头温煦的灯光照人入眠，在这清净隽秀的卧室中沉睡是最好不过的享受。

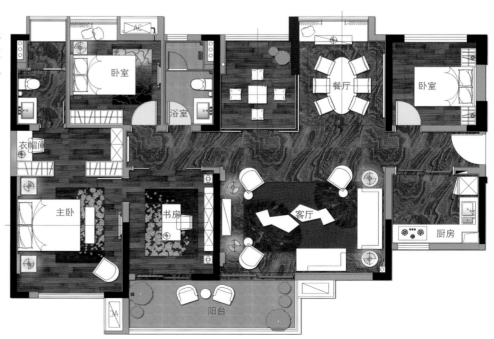

深沉悠远

是你宽阔博大的心胸

蕴藏着大自然的力量

宁静 祥和

浓如夜空的蓝 稳重柔和

淡如青烟的蓝 透明轻快

深远悠长间无尽遐想

BLUE 蓝

包容中式 幸福之家
INCLUSIVE CHINESE STYLE, HAPPY HOME

>>> 设计理念 | DESIGN CONCEPT

"Beauty is a common pleasant feeling" is the design concept which ZESTART has always insisted. Different from the Zen and tranquility of many Chinese show flat, ZESTART thinks that Chinese style should serve residents' lives so that they can feel pleasure and happiness from it rather than go to the alley of style for the sake of Chinese style. This project focuses on this kind of concept. The designer uses bold, international and fashionable technique to present purchasers' double needs of artistic tastes and life emotions, to break traditional Chinese layout to make it closer to internationality and to present a decent and inclusive Chinese flavor.

"美是普遍的愉悦感受"是则灵艺术长期坚持的设计理念。区别于大部分中式样板房的禅意与宁静，则灵认为即使是中式，也应该服务于居者的生活，使人能够从中体会到愉悦感和幸福感，而非为了中式而中式，走到风格主义的胡同里去。本案则集中体现了这种理念，设计师用大胆的、国际化的、潮流的手法表达了现在的购房人对于艺术品味与生活情感的双重需求，打破传统中式的格局，让其与国际接轨，更体现出一种大气包容的中式风味。

项目名称：昆明滇池龙岸云玺大宅十三区中式户型别墅样板间
设计公司：深圳市则灵文化艺术有限公司
设计师：罗玉立
项目地点：云南昆明
项目面积：873m²
主要材料：实木、布艺、陶瓷等
摄影师：陈中

>>> 会客厅：惬意的休闲

　　走进客厅便能感受到其散发出来的休息和惬意，白色的木质天花带着乡村风格的韵味，浅灰色的沙发线条明朗，白色大理石和蓝色斗柜上摆放这中式特色的塔和佛头。明目蓝色的书架成为空间的背景，深蓝灰的地毯和沙发蓝色抱枕为之呼应，木质的休闲，布艺沙发的惬意，让人能在其中尽情放松。

>>> 餐厅：宫廷盛宴

　　如同宫殿般的餐厅以双层挑高的设计体现出它的正式、讲究和独特的仪式感。方格天花以蓝色彩绘为装饰，呼应着窗边墙面背景色。能容下14个人坐位的餐桌以白色桌布装饰着典雅，两把紫色的主餐椅和12把水蓝色的副椅主次分明，黄色的摆件装点在餐桌上，靠窗一架古典钢琴随时可为就餐增添一首悠扬的旋律。

主体色
PRIMARY COLOR

点缀色
INTERSPERSED COLOR

>>> 色彩分析 | COLOR ANALYSIS

Blue is the main character to link the space, tranquil, eternal and intriguing. As the interspersed color, bright yellow is nifty, lovely and eye-catching. Proper soft decorations and modern expression of classical Chinese style make people in it feel the joy and happiness from the deepest heart. Self-examination and Zen thoughts are no doubt important, but life needs love and joy most.

蓝色是串联空间的主角色，宁静永恒，耐人寻味。明黄作为点缀色，俏皮活泼，醒目亮眼。恰到好处的软装配饰，古典中式的现代化表达，让身居其中的人从内心深处体味到愉悦与幸福。自省与禅思固然重要，然而生活更需要爱与欢笑。

>>> 次卧：文雅岁月

相比主卧的正式，次卧则显得更加休闲。不同蓝色在空间的运用，呼应了各空间的主色。它可以是青蓝灰的墙面、亮蓝色的床头矮架、深蓝绿的地毯花，不同深浅的蓝色运用在空间不同之处，也显示出和谐的美感。木色的长椅、灰墨色的床品文雅中又带着岁月之感。

>>> 主卧：古典中正之美

主卧宽敞舒适，容得下休息与休闲。做旧金属色的床架勾勒出中式建筑的轮廓，传递东方的中正之美。蓝色背景墙和蓝色沙发在色彩上递增，显示出层次感。床尾营造出的休闲品茶区有着中式的韵味，如白色禅意插花、绿色坐凳、深色改良的太师单椅，同时黄色的斗柜和蓝色彩绘山水画又让人眼前一亮。

一次关于文化与信仰的探索
AN EXPLORATION ABOUT CULTURE AND FAITH

>>> 设计理念 | DESIGN CONCEPT

It is deeply believed that the most significant meaning of culture is to pass down the original memory shared by human beings from generation to generation while transcending times. In respect of Yinma Valley by the Great Wall, facing the constant and profound historical accumulation, certain actions are requested to recall the "memory of field and culture" embedded in the greatest architectural works in the history of human civilization. The application of the dialogue between old and new embodying the ancient Eastern culture shows Chinese people's confidence on lifestyle and aesthetic appreciation, vitalizes the space and endows the space with rich and deep culture. The modern design language applied in this project, the artful play of peony, blue and white porcelain, traditional embroidery, batik fabrics, bamboo weaved pendant lamp and so on, demonstrate the deeply-embedded mighty life in real sense.

我们深信,文化最重要的意义,就是在于把人们共有的原始记忆,超越时代流传下去。就长城脚下饮马川而言,要在长城跟下,面对历史绵延不断的积淀,采取某种方式来对应这个人类文明史上最伟大的建筑工程,唤醒其累积的"场域和文化记忆"。我们选择用新与旧的对话,展示东方古老文化,表达中国人对生活方式的自信,对审美的自信,让场域活性化并带给空间丰富的文化深度。本案中现代设计语言的应用,抛开了小家碧玉的羞涩,牡丹花,青花瓷,老绣片,蜡染布料,竹编吊灯等元素的碰撞,真正体现出一种老而不古的磅礴生命力。

项目名称:长城脚下饮马川
软装设计:布鲁盟室内设计有限公司
设计师:邦邦、田良伟、邹珊珊
项目地点:北京
项目面积:120m²
主要材料:Kenneth Cobonpue,Fandango挂灯,Obra Cebuana 手工茶几等

>>> 色彩分析 | COLOR ANALYSIS

In the process of design, we also examine ourselves from another perspective, aiming to achieve the resonance of old and new through the approach of deconstructing culture and trend, breaking the routined life and bringing out the spark of creativity and inspiration. In here, you can not only find the Rongjiang hand made Aizen blue fabric sofa and carpet and wood color exposed beam frames, but also the natural tree fossil slices, natural travertines and red furnishings. Just as a set of works by Italian ceramic artist Paola Paronetto displayed on the TV cabinet, the artful integration of corrugated paper texture into ceramic works, in other words of using one material to highlight the other, may bring out an unexpected result.

设计中，我们也用另一个角度审视自己，解构文化与潮流，打破常规的生活状态，碰撞创意与灵感，实现新与旧的合奏共鸣。在这里，你不仅能找到榕江手织布蓝色染布料的沙发地毯、木色的外露横梁框架，也可看到天然树化石切片、天然洞石和红色家具摆件。

正如电视柜上意大利陶瓷艺术家 Paola Paronetto 的一组作品，将瓦楞纸肌理巧妙融入陶艺作品中，用一种材质来表达另外一种材质，得到意想不到的效果。

>>> 客厅：返璞归真

双层空间客厅返璞归真的设计以实木梁柱为支撑，形如灯笼的红色吊灯传递出中国红的喜庆，也契合这长城脚下古朴的空间。沙发以蓝色为主，拼接着红橙色和棉麻色，深蓝色的地毯上"开出"缤纷的花，簇拥繁华。圆形透明玻璃茶几，红色点缀的盆景于细节处增添了空间中式韵味的精彩。

>>> 餐厅：品味往事

人们围坐在原木色古朴的长桌，坐在棉帆线编织的正红色和亚麻色的椅子上，享受美味的晚餐，畅谈趣事，舒适又不失仪式感。灰色墙面上以一幅水墨油彩画为装饰，抽象中可品阅出千百种解读，是人生，是生活，是世间千千万万往事。

主体色
PRIMARY COLOR

点缀色
INTERSPERSED COLOR

>>> 阅读区：诗意生活

厚重洞石与薄钢板组合的书架，别具一种节奏感与戏剧性。棕黄色矮桌上放置着清新雅致的蓝色点缀茶具，正蓝色花瓶中插着芦苇草，静读几页书，闲吃一杯茶，坐在红色懒人椅享受冬日里的暖阳，"蒹葭苍苍，白露为霜"，抬眼望去，潮河边高高的芦花摇动，折几枝插在瓶内，瘦瘦的筋骨把生命的诗意一缕缕地挑亮，展现不一样的中式精彩。

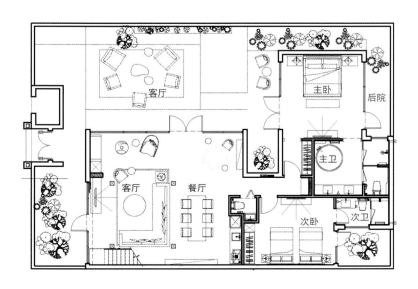

>>> 卧室：追忆似水年华

　　双人床的卧室设计似乎更为贴心，深湖蓝色的床品有着岁月沉淀下的静谧之感，给人以舒适和宽心。墙面上吊挂着白色纸片的装饰，不经意间让人回想到记忆中遥远的童年，蓝灰色的衣柜又带着年代岁月感。或回忆童年，或追忆往事，或体味沧老岁月，这里将能带你走进时间的长河。

三代同乐 典雅居所
ELEGANT RESIDENCE WITH JOYS OF THREE GENERATIONS

>>> 设计理念 | DESIGN CONCEPT

Wuhan Dongyuan Napa Valley Duplex Villa Show Flat is located in the world-famous Wuhan Optical Valley High Science and Technology Park and the targeted clients are wealthy and wise classes in Wuhan. ZESTART design team draws inspiration from nature to make tranquil French flavor penetrate in every interior corner and to wake up the long-lost peaceful, comfortable and pleasant feelings. The upper three floors of the whole villa are living places for the three generations and are designed according to residents' different life needs. The negative floor is the place which presents value of the villa. It uses imitated natural sky light to make the entire room bright, comfortable, natural and beautiful as if there are natural lights. The walls use natural rubbles whose delicate and rough textures bring people impacts and affects.

武汉东原纳帕溪谷双拼别墅样板间位于举世闻名的武汉光谷高科技园区内，目标客户群是武汉财智阶层。则灵艺术的设计团队从自然中汲取灵感，让恬静的法式气息渗透在室内的每一个角落，唤起人心中久违的宁静、舒适与愉悦之感。整栋别墅楼上三层作为三代同堂的生活起居之地，根据居者不同的生活需求进行了针对性设计。负一层则是别墅价值体现之地，采用仿自然光天灯，使整个房间犹如有天光洒入，明亮舒适，自然美好。墙壁用自然毛石，精细与粗犷的机理碰撞带给人冲击和感动。

项目名称：武汉东原纳帕溪谷双拼别墅样板间
设计公司：深圳市则灵文化艺术有限公司
设计师：罗玉立
项目地点：湖北武汉
项目面积：493m²
主要材料：实木、皮革、石材、布艺等
摄影师：曾康辉、黄书颖

>>> 客厅：蓝色魅力 灵动优雅

优雅的法式从来都是不惧任何色彩的装饰，水蓝色的墙壁以淡雅的基调退居空间后，显示出长条蓝色沙发的醒目，正蓝色的沙发线条柔美，宽大舒适，绿色、黄色、蓝色的抱枕点缀其中，活跃不失浪漫。铜黄色的弯曲蜡烛式吊灯，尽情展示着法式风格的典雅。

>>> 餐厅：中西相遇 味蕾时光

水蓝色的墙面萦绕四周，圆形深褐色的餐桌带着东方韵味，餐椅方正文雅，蓝色花瓶中插放着开得正好的鲜花，清香环绕。靠窗而放的黄色斗柜，两盏高脚台灯纤细雅致，蓝色灯帽与窗帘边相呼应，有阳光的日子里，在此进餐真是别样的享受。

主体色
PRIMARY COLOR

点缀色
INTERSPERSED COLOR

>>> 色彩分析 | COLOR ANALYSIS

The project sets elegant and decent blue as the main tone to link every space, supplemented by large piece of handmade natural colored drawings, creating a mysterious romantic breath. At the same time, dreamlike pink, mysterious coffee and dazzling yellow make the space send out a beauty of balance through the intersperses of different colors. Sun lights sparkle into interior, scattered and mottled, making the colors softer and more charming.

空间以淡雅大气的蓝色为基调串联了各个空间，铺以大幅的手工自然彩绘，营造出神秘的浪漫气息。同时加入梦幻的粉色、神秘的咖色和醒目的黄色，更让空间在各个色彩的点缀中散发出平衡之美。阳光洒落室内，错落斑驳，让色彩显得更加轻柔迷人。

>>> 卧室：和谐之美 温馨之居

法式风情独特的圆形纹理天花设计搭配带帽的吊灯，营造出温馨浪漫的氛围。红色背景墙醒目亮眼，粉色床品又俏皮可爱，蓝色地毯则打破了一片红色的暖，为空间加入理智的中性色调，让人在色彩均衡间找到空间散发出的最舒适的居住气息。

>>> 长者房：云中仙鹤 益寿延年

老人房以双人床设计为年迈的老人提供更舒适的晚年生活，深蓝色的床头背景墙和蓝色床品、地毯相呼应，相比浅色的水蓝，深色的蓝更有利于老人安神静气。卧室同时以鹤的吉祥寓意来装饰，如床头背景墙上云中仙鹤的金色挂件、床尾墙面深咖的斗柜和壁纸，柜子上站立鹤的摆件等，都用来表达期望老人长寿的美好愿景。

邂逅法式风情
ENCOUNTERING FRENCH STYLE

>>> 设计理念 | DESIGN CONCEPT

People are yearning for the magnificence, romance and luxuries presented by French style for a long time, but traditional French style tends to be heavy, gorgeous and complicated. So the designer intentionally abandons the complicated decorations in baroque period, stresses the combination of comfortable furniture and modern style, integrates French luxury and romantic elements with dignity and elegance of Oriental culture, uses different colors, materials and arts and crafts in different periods and presents them through delicate and rich details to make the entire space luxurious. The details contain a trace of romantic feeling; inadvertently, the decorative floral patterns reveal the romantic feeling featured by France.

法式风格传递出来的大气、浪漫和奢华令许多人向往不已，但传统法式往往厚重华丽繁复，因此设计师特意摒弃了巴洛克时期的繁复装饰，强调家具的舒适度并与现代风格相结合，将法式奢华、浪漫的元素与东方文化的端庄、典雅相融合，运用不同的色彩、材质，不同时期的艺术与工艺，通过精致而丰富的细节呈现出来，使整个空间奢华有度，细节中蕴含着一丝浪漫的情愫，不经意间，一处处轻描淡写的花纹流露出法国特有的浪漫情怀。

项目名称：西宁绿地云香郡别墅
设计公司：广州汉意堂国际软装
设计师：袁旺娥
项目地点：青海西宁
项目面积：245m²

>>> 客厅：不期而遇的浪漫

有着浓郁法式风情的客厅，以米白色的长条沙发和贵妃椅为焦点，讲究的布艺工坊给沙发更多的舒适，蓝色的靠枕与印花蓝色地毯相呼应，一上一下，为之灵动。沙发以金色收边雕刻其轮廓，红褐色的茶几上鲜花簇拥，法式茶具雅致，都是法式风格在空间中具体的展现。

主体色 PRIMARY COLOR	
点缀色 INTERSPERSED COLOR	

>>> 色彩分析 | COLOR ANALYSIS

French elegance and romance need reasonable color collocations to present its flavors. Based on white and creamy white, interspersed with blue furniture and gold edging, jazz white marble, blue printed cloth and different flowers and furnishings add some highlights for the space; in particular, as main tone or interspersed color, blue which links spaces makes the space harmonious and unified.

法式的优雅浪漫一定是需要合理的色彩搭配方能尽显其味，空间以白色和米白色为肤，蓝色家具和金色收边为点缀，爵士白的大理石，蓝色印花布艺，各色的插花摆件都为空间增添不少亮点，特别是用来串联空间的蓝色，以主角色或点缀色让空间整体上混为一体，和谐统一。

>>> 餐厅：典雅中的仪式空间

双层挑高的客厅以其大气奢华彰显着屋主对就餐的重视，白色略带墨迹的地砖和米黄色印迹壁纸，极力营造出大气洁净的就餐空间，垂直而下的窗帘以蓝色为帷幔，呼应着餐椅坐垫。餐桌椅有着讲究的金色轮廓，金色的桌脚上有细致的雕刻，齐全的餐具和芬香的插花，为整个空间增添不少就餐的正式之感。

>>> 主卧：温馨梦乡

有阳光的时候，拉开窗帘在主卧能尽享休息和阅读的趣味。米黄色的床品充满了轻柔的质感，床品上淡雅的蓝色床毯和窗帘不约而同地相遇。黄褐色的书桌和地砖相呼应，静雅的黄褐色给人沉淀心灵的感觉，更适合主人在此阅读办公。

>>> 男孩房：飞扬青春

所有用来形容男孩的特质都能在他的卧房中找到，可以是摇滚的黑色吉他，彰显个性的深蓝色挂画，也可以是青春活力的蓝色床品，并以其他色彩抱枕为之点缀。虽然男孩房不够宽敞，但带有独立的阳台，也足够装得下他青春中的休息活动和幻想。

浪漫爱琴海
ROMANTIC AEGEAN SEA

>>> 设计理念 | DESIGN CONCEPT

When it comes to romance, what you think of is French romance, walking by the Seine, wandering under the tree shades in Champs Elysees, drinking in the bar and whispering in the coffee. Mr. Liang Sicheng once said: "A good designer needs the mind of philosophy, sight of sociologist, accuracy and practice of engineer, sensitivity of psychologist, insight of litterateur and expressiveness of artist." In designs of this project, the designers use "not following" design technique to present romantic Aegean Sea design style, combine with modern people's yearning for concise yet not simple life and inject elegant and luxurious European furnishings into the space. Chic furniture modeling, textured cloth supplies, glittering and translucent utensils and fashionable and mellow lines, all these appropriately present the ingenuity and originality of the designers.

提及浪漫，人们第一联想到的就是法国的浪漫：塞纳河边的闲庭信步，香榭里舍树荫下的低头徘徊，酒吧里的慢饮斟酌，咖啡馆里的窃窃私语。梁思成曾说："一个好的设计师必须要有哲学家的头脑、社会学家的眼光、工程师的精确和实践、心理学家的敏感、文学家的洞察力和艺术家的表现力。"在本案的设计中，设计师以"不跟随"的设计手法让浪漫的爱琴海设计风格表现力十足，结合当下人们对于简约而不简单的生活的向往，注入了欧式家居的典雅和华贵、别致的家具造型、触感十足的织布用品、晶莹剔透的器皿、时尚婉转的线条，这些都恰到好处的表现了设计师的匠心独运，独领风骚。

项目名称：天池公馆270A户型样板房
设计公司：奥迅室内设计
主创设计师：罗海峰
参与设计师：罗皓、朱芷谊团队
项目地点：云南昆明
项目面积：415m²
主要材料：维纳斯灰石材、天空蓝石材、鱼肚白石材等

主体色
PRIMARY COLOR

点缀色
INTERSPERSED COLOR

>>> 色彩分析 ｜ COLOR ANALYSIS

"The Aegean Sea is the place which is closest to heaven." In designs of Tianchi Apartment House Type 270, soft and elegant blue white makes the entire space fresh and cool. The designers bring the brush of blue in the sea shore into the whole interior space, manifesting a warm atmosphere, bright and amiable. Dreamlike blue is used into extreme; appropriate white adds some romantic feelings for the whole space.

"爱琴海是最接近天堂的地方"，在天池公寓270户型的设计中，柔和高雅的蓝白色调让整个空间倍感清新爽朗，设计师将海岸上的那一抹蓝延用到室内的整个空间，显示出温馨之气，明亮而又亲切。梦幻的蓝色被发挥到了极致，恰如其分的灰白，让整个空间增添了几分浪漫情怀。

>>> 客厅：蓝色风情

以米白色沙发作为基调，用蓝色加以升华和点缀，墙上的艺术挂画、电视机柜上的装饰盒、桌上深海蓝玻璃器皿非常应景，打破沉睡中的寂静，让设计的主题在此得到了深化，浅水蓝色的软榻让人感觉坐在柔软的沙滩上，把蓝天白云下的爱琴海引向空间，让人近距离欣赏美景。

>>> 红酒雪茄室：品味诗和远方

位于负一层的红酒和雪茄室作为主人的秘密花园，并没有过于繁华的装饰，更多的是返璞归真。蓝色抱枕和深蓝色的地毯有着色彩上的递进，沙发后面的梅花枝艺术屏风也许就是主人的人生写照，"生活不只眼前的苟且，还有诗和远方"。

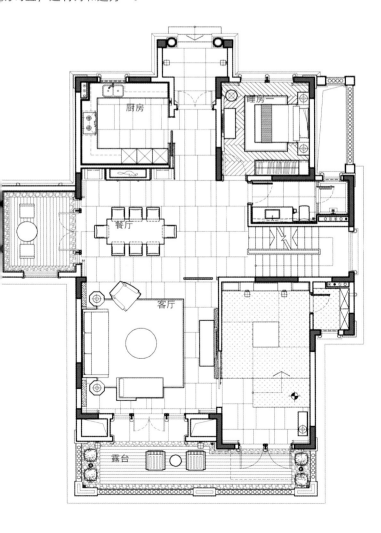

>>> 主卫：沐浴时光

　　主卫地砖选用冷色系的维纳斯灰大理石和墙壁的鱼肚白大理石有异曲同工之妙，如海水洗刷过的沙滩，留下了些许斑迹，配合屋顶婉转流畅的赤金色艺术吊灯和洗脸池旁边的蓝色花瓶插花，不自觉地增加了空间的视觉冲击力。

>> 主卧：静卧舒室

　　拾级而上，二楼的主人房让人顿时眼前一亮，白色立体天花和深蓝色盛开黄花的地毯再次响应蓝天、白云和海的意象设计，搭配米白色的家具和床品，大气十足。床头挂画如风掀起的蓝色波浪，爱琴海波涛的回声似乎在耳畔响起。

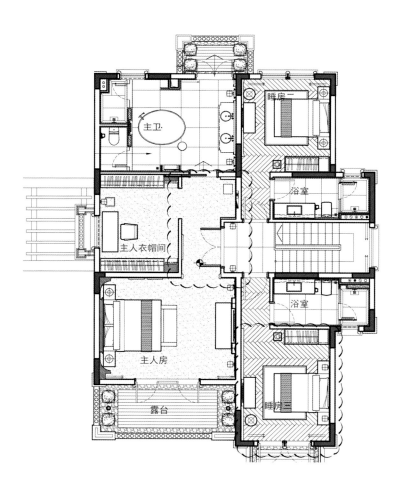

绅蓝空间
GENTLE BLUE SPACE

>>> 设计理念 | DESIGN CONCEPT

We live in this colorful world and what we see first is color. The designer especially needs to understand meanings of works endowed by colors, fully uses color changes to stimulate the audiences' senses and emotions and leads them to dig out meanings behind colors. Interior color collocations especially need to develop vigor and tension of colors. As for absolute control and creativity of "color", excellent designers need to be fond of "color" and then they can make a perfect design. The designer uses profound aesthetic accomplishments to combine decorative elements with contemporary designs, uses colors to express the understanding of space and uses colors to endow the space with temperament to create an international British modern style.

  我们生活在这个缤纷的世界里，触目所及的第一焦点就是色彩。而设计师尤其要读懂色彩赋予作品的意义，充分运用色彩的变化来刺激受众的感官和情绪，引导他们挖掘色彩背后的意义。室内空间的色调搭配，尤其需要发挥色彩的活力与张力，对于"色"的绝对把控和创造力，优秀的设计师唯有好"色"，才能把设计做好。设计师以深厚的美学素养，将装饰元素结合当代设计，用色彩表达对空间的理解，用色彩赋予空间的气质，开创了国际范的英伦摩登风格。

设计公司：金元门设计公司
设计师：葛晓彪
项目地点：浙江温州
项目面积：200m²
主要材料：木地板、涂料、玻璃等
摄影师：刘鹰

主体色
PRIMARY COLOR

点缀色
INTERSPERSED COLOR

>>> 色彩分析 | COLOR ANALYSIS

As a painter, the designer can easily control the colors. The designer boldly uses colors and sets royal blue with elegant and romantic gentleman aristocratic temperament as main tone of the space, which is as beautiful as immersing in the endless and tranquil lake water. Royal blue which almost fills all spaces is surrounded by yellow retro oak grain floor; cold and hot, it is the collision of sense and passion. All kinds of dark and light blue colors in different shades pave in the space and endow the space with the most unique blue space temperament.

  身为画家的设计师，对于色彩的拿捏自然驾轻就熟。设计师使用色彩的手法新奇大胆，在本案中注入优雅浪漫的绅士贵族气质的皇家蓝作为空间的主色调，美丽的像是沉浸在无尽的静谧的湖水中。几乎充满整个空间的皇家蓝包裹着黄色复古橡木纹的地板，一冷一热，是理智和激情的碰撞。各种深浅浓淡的蓝在空间中铺陈开来，赋予空间最独一无二的蓝调空间气质。

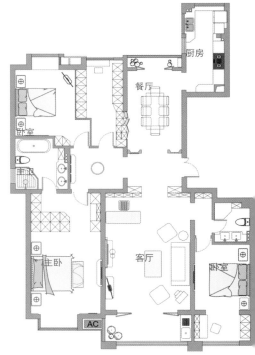

>>> 客厅：蓝色森林

客厅无疑是本案的剧情高潮，几乎盈满整室的蓝，从天花板到背景墙，从书柜到地毯，蓝得肆意，蓝得放纵，蓝得极致，具有十分强烈的画面感和视觉冲击力。蓝色的运用在这里达到顶峰，让时间凝固，场景定格。

>>> 沙发及背景：春波含绿

墨绿的沙发，几乎淹没在蓝色背景的客厅中。但你无法忽略它的存在，如此浓烈张扬的绿，带着魅惑的光泽与质感，于一片蓝色森林中绽放炫目光彩。色泽明亮华丽的金色镜框及画框，是设计师的点睛之笔，仿佛清冷冬日里的一缕阳光，明媚而炽热，瞬间点亮整个空间。

>>> 餐厅：浪漫交响曲

金色的叶子灯饰脉络清晰可见，纹理鲜明，是出人意料的小惊喜。设计师将皇家蓝的绅士气质和灰度粉的浪漫雅致，完美地融入餐厅空间。粉绿间隔摆放的座椅，红蓝对立的墙面，色彩交错，共同营造了优雅迷人的进餐空间。

>>> 主卧：梦里花落知多少

干净明亮的白色天花板犹如冬日初雪，带来清冽的自然气息。渐变蓝色则铺满了整个床头背景墙，凸起的格子装饰带来立体的视觉感受。单人沙发、枕头、镜框、地毯，都染上了欲说还休的蓝，优雅而宁静，理性而浪漫。

住在米兰的鬼马小姐
MISS FAIRY IN MILAN

>>> 设计理念 | DESIGN CONCEPT

The pencil quietly draws the sketch; the date is written in the lower right corner of the canvas; who can believe that a drop of rain can become a cup of coffee? The bright red in the face turns into rose. Miss Fairy is a little magic and only loves the brush of innocence in this mature world.

Hands of the illustrator seems to have magic and can turn vegetables into fairytales in the dining table. Mature avocado plays with fresh lemon; the natural salt of wheatgrass joins into Swiss chard; these are the favorite light dishes of Miss Fairy who is infatuated with the taste of the food itself. Her most beautiful appearance is the gentleness to life.

A house, two people, three meals and four seasons, they together enjoy five tastes of life. Miss Fairy says her favorite work is the time she spent.

　　铅笔安静地画着素描，年月写在画布的右下角，谁会相信雨滴会变成一杯咖啡？脸上的绯红幻化成玫瑰。鬼马小姐有点神奇，在这成熟的世界，独爱一抹天真。

　　插画师的手似有魔力，蔬菜会变成餐桌上的童话。熟好的牛油果撩拨着柠檬的清新，冰草的自然咸味闯入瑞士甜菜，这是鬼马小姐最爱的轻食主义，迷恋食物本身的味道，她最美的样子，是对生活的温柔。

　　一房两人三餐四季，共飨生活五味，鬼马小姐说，我最好的作品，就是我度过的时光。

项目名称：南昌绿地朝阳中心精装户型样板间
软装设计：成象软装
空间设计：梁志天设计师有限公司（SLD）
项目地点：江西南昌

〉色彩分析 | COLOR ANALYSIS

the whole, the designer chooses concise tone and comfortable modeling, follows the principle t "highlights can be less, but if there is highlight, it must light the life" and manifests quality and ndation. Warm white and profound blue are used throughout the entire space; delicate and elegant blue he soul of the space; the layers are transparent and the colors and luster are bright and fresh.

设计师整体选择以简洁的色调，舒适的造型为主，遵循着"亮点不宜多，有即点亮生命"的原则，细微处彰显品质与功底。温馨的白色与深邃的蓝调为主调贯穿整个空间，细腻优雅的蓝，是空的灵魂。层次通透，色泽鲜明。

主体色
PRIMARY COLOR

点缀色
INTERSPERSED COLOR

>>> 客餐厅：格调序曲

　　悦耳的音乐在咖啡里苏醒，温柔是写给米兰的一首小情歌。天生的简约挚爱狂也有百变千面，简而不单。蓝调也亟需酷劲儿相随，一桌一椅都蕴藏着艺术家的小宇宙。

>>> 书房：魔法天地

　　飞鱼可上天，鸟兽能语言，只需拿起笔，这一方简约的工作台就是妙趣横生的魔法天地。深蓝与墨黑，都可以随你的心情任性飞舞。

>>> 卧室：与爱相拥

摇曳时，微妙变，酣睡时，水墨蓝，演绎浪漫。跳跃的金色线条精致而立体，与咖色的皮革床具相宜得章。而次卧室里毛茸茸的毯子野性而温暖，让空间肆无忌惮地拥抱独特。

海洋坐标去旅行
TRAVEL ON THE OCEAN

>>> 设计理念 | DESIGN CONCEPT

Wandering in the home of beach and ocean, let's have a spiritual journey.

No matter life or livelihood, what we pursue is a natural and harmonious beauty. Life is about choice. Placing yourself in a proper position and choosing the suitable life and lifestyle, as long as the heart is happy, it is the best.

The designers plan the entire space by this kind of life attitude and choose fresh orange which represents joy and vividness and makes you think of energies. This kind of color is not only warm but also full of happiness and hope. Using it in home design, no matter an item or the whole tone, fresh orange can make your home lively and beautiful forever.

漫步在沙滩与海洋的家，来一次心灵的旅行。

人生也好，生活也罢，我们追求的是一种自然与协调的美。生活，在于选择。把自己摆在一个合适的位置，选择适合自己的生活与生存的方式，只要心是快乐的，那就是最好的……

设计师以这样一种生活态度来规划整个空间，选用新鲜的橙色，代表着欢快活泼，看到橙色就会联想到满满的能量，这样的色彩不仅温暖还充满着幸福与希望。把它用于家居设计，无论是一件单品还是整个色调，鲜橙色都将使你的家充满活力，靓丽永久！

项目名称：杭州阳光郡D户
设计公司：西盛建筑设计（上海）事务所
设 计 师：何文哲、唐瑶华、熊成、王书璐
项目地点：浙江杭州
项目面积：175m²
摄影师：朱沈锋
主要材料：木地板、壁纸、布艺等

>>> 色彩分析 | COLOR ANALYSIS

With navy blue as the background color of the space, the color with magic makes the whole space dynamic. Vibrant orange is partially dotted, making the space vision different in size; they echo and attract each other, forming a visual sense of aesthetics.

　　定义海蓝色为空间背景色，一种拥有神奇魔力的色彩，让整个空间焕发生机。局部巧用活力橙的装饰点缀，让整个空间视觉大不一样。相互照应、彼此吸引，形成视觉上的层次美感。

>>> 客厅：诗意的栖居

优雅柔和的海蓝色背景仿佛风和日丽时的湖水，倒映着天空与白云，波澜不惊。它为家注入诗意绵长的气质和无比轻柔的质地，枝桠的吊灯、戏剧的脸庞、柔软的沙发，仿佛一切都在编织一个永不结束的唯美梦境。

>>> 挂画：生活艺术

有花，有鸟，有生活。浪漫而温馨的木质白条，热烈而欢脱的橙，精致灵巧的挂件，在一抹孔雀蓝的辉映下，更显情调与艺术。

>>> 餐厅：玩转时尚

延续着客厅的清新亮丽，玩转色彩的激情碰撞，同时结合民俗布艺产品和民族风格特色的装饰物，独具情调，显露韵味之余还兼备时尚之感。

>>> 主卧：对比显风情

初感仿佛是灿烂如洗的天空，蓝的净透。而火热深邃的橙既令人惊喜又饱含成熟风韵。以蓝为主导，橙充当点亮空间的暖色，互补前进，从而引起人们的注意力并激发热情，整个空间呈现出活泼愉悦的氛围。

朴素随和
是你充满内涵的沉默
在各大场景中
谦恭 寂寞
遇鲜艳的暖色 温和典雅
遇较纯的冷色 高贵热情
古典中性间沉稳奢华

G R A Y

灰

传承人文新风范
INHERITING HUMANISTIC NEW STYLE

>>> 设计理念 ｜ DESIGN CONCEPT

In the design of Shoucheng Longxi, LSDCASA uses modern design technique to deduce and arrange these traditional elements, extracts warm color and luster from jade and porcelain, pink gold and dark black from palace silk scroll painting to form the main tone and interspersed colors and skillfully turns heavy dignity of copper and nobility and softness of mercerized velvet into comfort and practicability of the space so as to naturally reveal Chinese power and fun and to make the "home" concept of Chinese mansion return to the pursuit of the comfort of life.

在首城珑玺的设计中，LSDCASA以现代的设计手法，将这些传统风物元素进行演绎和加工，将玉石、瓷器的温润色泽、宫廷绢本画的赤金、玄色等提取出来形成主色调和点缀色调，将铜器的厚重威仪和丝光绒的贵气柔软等材质，巧妙转化为空间之间刚柔并济的舒适实用，借此，自然地将中式的力量和意趣呈现流露，将中式豪宅中对"家"的概念真正回归到对生活舒适性的追求之中。

项目名称：北京首城珑玺样板间
设计公司：LSDCASA
设计团队：LSDCASA一部
项目地点：北京

主体色 PRIMARY COLOR	
点缀色 INTERSPERSED COLOR	

>>> 色彩分析 | COLOR ANALYSIS

Space design of this project absorbs Chinese concepts and elements at the same time uses modern design language to express living demands. The concept which stresses symmetry layout and the technique which is good at neutral colors in Chinese style are presented in this project. No matter large pieces of cotton gray and landscape splash ink color or interspersed and supporting pink gold and dark green and the landscaping charcoal gray, all shine decent and fair manner of Chinese colors. Modern materials match with gold elements and dark and black carpet, making the whole space archaistic with modern imprints.

本案空间设计汲取中式理念和元素，同时采用现代的设计语言表达居住诉求。东方中讲究对称布局的理念和善用中性色彩的手法，在本案中得到体现。无论是大片的棉麻灰、山水泼墨色，还是作为点缀和配角的赤金黛绿色、善用造景的炭灰色，都闪耀出中式色彩的中正之气。现代材质的运用加入金色元素的点缀和黑白相间的地毯，让整个空间既仿古又不脱离时代的印迹。

>>> 客厅：温柔岁月

环形多层吊灯让双层挑高的客厅更具层次感，棉麻灰的墙体环绕四周，似乎诉说着岁月沉淀中的美好故事。背景墙上传统中国式的建筑挂画，以岩石蓝打底，三五孩提用彩色点缀，活跃其中。客厅整体引入中式对称之美，中轴排开，主次分明，韵律与秩序同在，隐藏在形式背后的是当下对传统人文精神的敬畏与继承。

>>> 餐厅：水墨山色

"山"在中国水墨画中是最重要的意象之一，山形的水晶吊灯则是用现代材质描摹水墨山水的意境。淡墨色的地毯延续了"山"的元素，透过丝绒的材质，"山"在这里显得气韵万千，既像山，又像云。雅致灰色的餐桌椅融入空间，餐桌之上选用红橙色的石榴代替花束，别致而活泼。

>>> 品酒区：把酒畅谈话乾坤

　　木色为背景的品酒区放置着深灰色桌椅，雅致的茶具轻放，翠绿的竹筒器皿清新点缀，配以枯枝插花，散发出静谧清幽的气息。每个男人都需要这样一个空间，容得下诗和酒，容得下三五知己，容得下这一路走来的风与月。男主人儒雅绅士又兼具摩登时尚的生活情趣，在这里一览无遗。

>> 主卧：艺术东方

床头背景灵感来源于日本江户时代重要艺术家 Tawaraya Sotats 绘制的扇面散屏风，中国水墨画融入画中的图案，做旧金箔墙纸打底，采用工笔画法，打造出一副融合了江户风貌和中国韵味的装饰墙纸。墨绿与赤金暗生款曲，金属与棉麻浅唱低吟，让空间在质朴雅致的意境中又分外地提炼出一丝当代气质与空间契合，不温不燥、不多不少、恰当刚好。

春风笑 润绿珠
SPRING BREEZE SMILES AND MOISTENS GREEN BEADS

>>> 设计理念 | DESIGN CONCEPT

Ba Shu Area is a land of abundance. The poet Du Fu who lived here because of the Rebellion of An Lushan and Shi Siming once wrote: "He died before he accomplished his career. How could heroes not wet their sleeves with tear on tear", "Good rain knows its time right" and "At dawn, we will see red flowers on the moist soil, and the town will be colorful." There are historical profoundness, King's openness, poetry's gentleness and Sichuan opera's clang. Holding a cup of tea and arranging troops according to the Dragongate tactics, the royal court is inside, so are the universe and Vanke Emerald Park Townhouse Show Flat designed by Creative Space. Here, historical context and comfortable and elegant life feelings converge together and deduce an enjoyable scroll of contemporary Chinese lives.

The villa has a total of five layers. According to the living requirements of three generations, the designers plan reasonable activity zonings and create two high courtyards under the ground and between the first and second floors on the ground respectively to enhance penetration and flow of different layers, to keep relaxing, bright and transparent space feelings and to lay a foundation for designers to create a living aesthetics.

蜀地，天府之国。安史之乱流落此地的杜甫曾写："出师未捷身先死，长使英雄泪满襟"，也写"好雨知时节"，"晓看红湿处，花重锦官城"。这里有历史的深邃，王者的开阔，诗词的温婉，川剧的铿锵。端一盏茶，摆起龙门阵，庙堂江湖在里头，天地万物在里头，创域设计出品的万科翡翠公园联排别墅样板间也在里头。在这里，悠久的文脉，舒适雅致的生活情怀汇聚在一起，演绎出一场当代中国人生活的写意画卷。

整栋别墅共有五层。设计师根据三代同堂的居住要求，划分出合理的动静分区。并在地下一层，地面一、二层之间分别打造出两个挑空中庭，以增强不同层面空间的穿透与流动，也保持轻松、明亮、通透的空间感受，为设计师营造整体的居住美感奠定了基础。

项目名称：成都万科翡翠公园别墅样板间
硬装设计：深圳创域设计有限公司
软装执行：殷艳明设计顾问有限公司
设计师：殷艳明
参与设计师：文嘉、万攀、周燕黎、周宇达、梁深祥
项目地点：四川成都
项目面积：450m²
主要材料：玉石、茶色不锈钢、皮革、树脂板、墙纸、橡木烟熏木地板、灰茶镜、透光云石灯片、艺术玻璃等

主体色
PRIMARY COLOR

点缀色
INTERSPERSED COLOR

>>> 地下一层：休闲之意

　　地下负一层也是本案的重点所在。不同功能空间的主题打造，提升了整个设计的品味和韵味。负一层增设夹层空间，下沉庭院、雅灰会客区、宝蓝台球室、影音室和棋牌室一体化设计体现多重娱乐功能。在设计上，具有与公共空间相同的表现主题，围聚起家人好友间的情感。

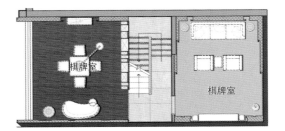
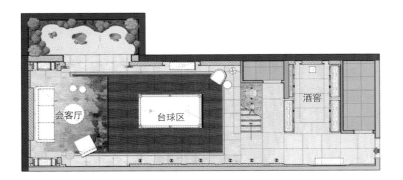

>>> 色彩分析 | COLOR ANALYSIS

The whole space combines warm gray with wood color to create a warm and elegant atmosphere. Partial stainless steel stresses the linear structure of the space; vermilion, pink gold, ultramarine, black gray and dark blue are used throughout every space to make the home vivid and lively in a sincere and easy calmness. Here, the beauty of scenery, the beauty of space, the beauty of artistic conception and the beauty of materials converge in details, obscure boundaries between each other, reach a permeability and reflect the essential spirit in the perspective of traditional aesthetics that "meaning is beyond form and the form is forgotten when the meaning is got" in modern people's lives.

整体空间以暖灰调结合原木色为主，营造温馨雅致的氛围。局部不锈钢强调空间线性结构，贯穿各个空间的朱红、赤金、群青、苍色、藏青点缀，让整个空间在一派敦厚宽容的沉静中，又生动活泼起来。在这里，山水之美、空间之美、意境之美、材质之美都在驻足回首之间，融汇贯通，消隐了彼此之间的界限，而达于通透，在现代人的生活中体现出"意胜于形，得意忘形"这传统美学意义上的本质精神。

>>> 客厅：浓墨重彩山水画

　　一边墨线恣意纵横，灰色扑洒中见山峦叠嶂，江河流淌，墨色层层晕染，似云烟，似雨雾，充盈于天地之间。这整面墙的水墨意境气势磅礴，自然天成。另一边将中国传统屏风的概念扩展以分割墙面，黄铜丝打造的莲叶与莲蓬错落有致地分布其间。两组主要的沙发一白一蓝，形式现代又点缀中国传统元素。

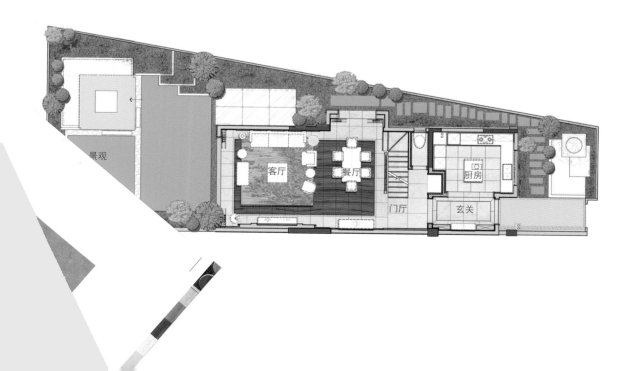

>>> 楼梯：蕙质兰心

　　楼梯处采用透空的设计手法，联系一、二层挑空的中庭。中庭设置餐厅，从高空轻盈垂落的艺术吊灯与不锈钢屏风营造出华丽的气质，又强调了中庭挑高的纵向视觉效果。餐桌上，一簇红兰在沉稳的花器上显得分外典雅，搭配蓝色的餐椅显示出女主人兰心蕙质的高洁之态。

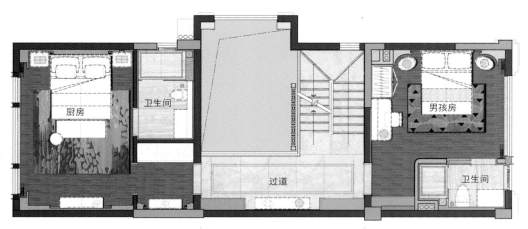

>>> 男孩房：飞扬的梦

男孩房的设计以"飞机"主题贯穿整个空间，灰色为底，蓝色点缀，相关饰品的搭配、穿插不仅体现了孩童时代大胆想象、探索，充满活泼动感的特质，也隐喻了《小王子》这本经典童话中对生命纯真、本质的永恒追求的美好愿望。

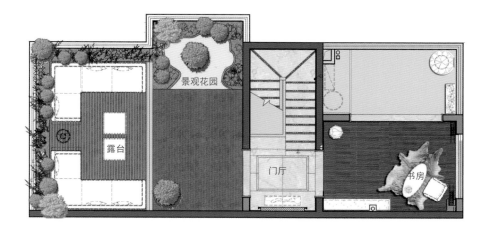

宁静港湾
TRANQUIL HARBOR

>>> 设计理念 | DESIGN CONCEPT

This project is a double duplex residence whose original structure is standard two living rooms and two halls. The layers are statically and dynamically divided; the first floor is mainly activity areas; the second floor has two bedrooms and it is designed into a work and living space according to the owner's requirements. The first floor is used for reception with kitchen, dining room, bathroom and chatting room. Living up to the outside river scenery, the designer brings the original balcony into interior to make the owner maximally enjoy the beautiful scenery given by nature. The second floor is mainly for rest with rest room, bathroom and study, with river scenery as well.

The designer symbolizes and abstracts modern and fashionable decorative languages to make them fit modern people's aesthetic concepts and make the environment primitive and elegant with fashionable sense.

本案是一套双层复式结构住宅，原结构为标准的两室两厅，动静分层，一层为主要的活动区域，二层是两间卧室，应业主要求将其设计为办公会所兼生活的一个空间。一层主要作为接待用，设有厨房、餐厅、卫生间、洽谈区。设计时考虑到为了不辜负窗外江景，便将原阳台纳入室内，让业主能够最大限度享受大自然给予的美景。二层主要为休息区域、休息室、卫生间、书房，同样江景也为室内所有。

设计师将现代时尚的装饰语汇加以符号化和抽象化，使之符合现代人的审美观念，使环境既古朴典雅，又不失时尚感。

项目名称：按蓝
设计公司：美迪装饰赵益平设计事务所
设计师：赵益平
项目地点：湖南长沙
项目面积：100m²
主要材料：定制木制品、进口黑色不锈钢、墙布、黑镜、清波、丝布刺绣、大理石、木地板等

主体色
PRIMARY COLOR

点缀色
INTERSPERSED COLOR

>>> 色彩分析 | COLOR ANALYSIS

The core of the whole design is to use blue as the bond to link the space and to express features of the space. Blue is a dreamlike color and gives people clear and romantic feeling. Under the foil of white, blue becomes fresher and more elegant, endows the space with meanings and makes the noisy heart bank in tranquil harbor. The bold application of gray blue makes people relaxed spontaneously.

　　整个设计的核心是通过大胆地运用蓝色作为连接空间的纽带和表达空间的特性，蓝色是一种梦幻般的色彩，给人清澈、浪漫的感觉。蓝色在白色的映衬之下显得更加清新淡雅，赋予空间意味，让喧嚣的心灵靠岸宁静的港湾。灰蓝为主色调的大胆运用，让人清爽的心情油然而生。

>>> 客厅：沉淀喧嚣的心

　　大块灰的使用有种低调神秘的美，让身处浮躁的都市生活中的人，在如此静谧的空间里释放压力，让喧嚣的心灵得到沉淀。钢架玻璃制作而成的电视背景墙，直通二层，既充当隔墙的角色，又为空间带来视觉上的通透感。同时设计师选择了不锈钢和黑镜能够让空间得以扩展，且硬朗时尚。

>>> 立体墙饰：花非花　蝶非蝶

　　特别的立体墙饰，最初意向为木质材质，但是最终为不锈钢喷漆完成，从一楼延伸至二楼慢慢绽放。美丽墙饰或花或蝶，由窗边一直延伸至楼梯的整个墙面。

>> 隔断门：隐匿美与魅

大面积蓝色的刺绣制作的隔断门分隔卧室与书房，如花语般富有诗意。"淡淡的幽蓝，隐匿着你的美；深深的一抹微蓝，是你的魅。"

>>> 木饰面：静谧幽蓝

空间主要木制品以尼斯木饰面本色为主，哑光低调让那一抹静谧的蓝色从中凸显而出，相互平衡。博古架上精致的不锈钢架与洞石相结合，为空间带来浓郁的人文艺术气息。

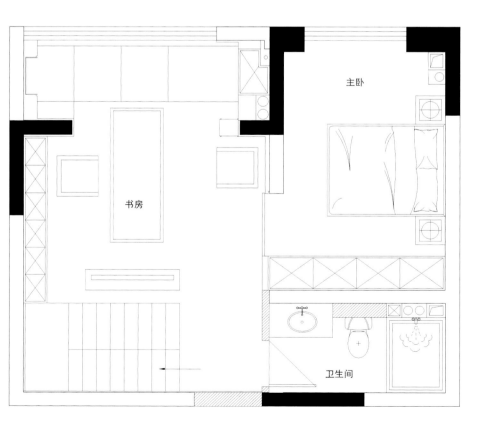

灰

素念禅心
PLAIN MIND AND ZEN HEART

>>> 设计理念 | DESIGN CONCEPT

The beauty of Hangzhou lies in West Lake, Xiling and Xixi. If living at the shore of West Lake is an extravagant hope in modern time, then living in the prosperous metropolis in Xixi Wetland, you can slightly taste the only remain of the most profound life flavors.

The designers take the meaning of "peace" from Oriental Zen, aiming to depict an association of "peace, color and form" as a whole. In the whole design, the designers turn them into atmosphere and artistic conception of the space so as to trigger residents to percept tacit resonation and pleasant process. In addition with modern design method and delicate material presentation, such as decent marble round table, edgings made of aluminum alloy materials, landscape painting background wall of colored glaze in blue and green, the designers use the prosperities of the materials themselves, trying to create an extremely elegant Oriental charm and considerate humane care and back to the most natural state of life.

The designers try to present us a home with the tranquility after getting rid of vanity and the existence of comforting distracting and complicated minds. It can infiltrate our minds inadvertently and become energies of our lives, which is as warm and tranquil, gentle and soft as the sunlight in winter.

项目名称：金地西溪风华洋房样板房
设计公司：杭州易和室内设计有限公司
设计师：麻景进、祝竞如、马辉
项目地点：浙江杭州
项目面积：132m²
主要材料：橡木饰面、意大利灰大理石、山水玉大理石、玫瑰金不锈钢、夹宣玻璃、绢画、硬包等
摄影师：阿光

杭州之美，不过三西，西湖、西泠和西溪。倘若，居于西湖之滨，在当代已然成为一种奢望，那么，西溪湿地，恰是繁华都会中，轻读杭城最深生活滋味的唯一遗存。

设计师取意东方禅之"静"，旨在描画"静、色、形"一体元素的联想，并在整体设计中，将其转化为空间的氛围意境，达到引发居者感知启悟默契共鸣和愉悦的过程。加以现代的设计手法与细腻的材质表现，如气派的大理石圆桌、铝合金材质的包边、琉璃材质蓝绿撞色的山水画背景墙等，更多地发挥材料本身的属性，力求造就绝代风华的东方神韵与体贴入微的人文关怀，回归到人最自然的生活状态中。

设计师想呈现给我们的家，可能是褪去浮华后的宁静，能抚慰纷扰繁杂的心绪所在。像冬日的阳光温馨恬静，和煦温柔，不经意间浸润我们的心田，成为滋养我们生命的能量。

主体色
PRIMARY COLOR

点缀色
INTERSPERSED COLOR

>>> 色彩分析 | COLOR ANALYSIS

The whole space sets elegant and clean gray as the keynote, with fresh blue interspersing in it; without stick ink and heavy colors, the designers explore the integrating point of design and feeling and creates a tranquil place to get away from trouble by powerful design language. Graceful times and elegant and cool breeze bring some feelings so that you can seek for the inner voice and back to Oriental leisure and elegance.

设计师不着浓墨，不加重彩，只用优雅干净的灰为基调，清新的蓝点缀其中，以简洁但有力的设计语言构建了一个安放纷扰的静谧之地。岁月雍容，雅韵清风，带着些许情怀，追寻内心的声音，回归东方的闲情雅致。

>>> 客厅：澄江明月

一进客厅，禅意东方的浓郁气息便迎面而来。以灰色为主色调穿插高质感的蓝，颜色清淡，有涤荡人心的效果。家具的造型简约，采用棉麻、胡桃木、金属等材质打造空间内的气质元素，与香道饰品完美融合，收放自如地诠释了东方的精髓，让人倍感上流社会的优雅与品味。

>>> 餐厅：山色缥缈

随着视线自然过渡到餐厅，餐厅的设计兼容了宴客礼仪和文人情怀，在灰色的基调中，璀璨的水晶吊灯洒下温馨的光景，设计师别出心裁地以写意山水画来装饰背景墙。气派的大理石圆桌、独家定制的餐椅与中国茶道在同一时空对话，隐喻独到的审美和非凡的气度。

>>> 书房：窗外千山绿

窗外的景色是书房的对景，视线毫无遮挡地从室内延伸至室外，空间也显得更加宽敞与明亮。设计师匠心独运，将窗外绿景也"纳入"室中，创造出全新的构景效果及审美体验。

>>> 主卧：月下竹声吟

主卧设计避免堆砌，寥寥几笔勾勒出沉稳大方的空间格调。干净清爽的灰搭配清新的蓝，组合出宁静禅意的空间氛围。原木纹理的家具，凸显着原始之美。于山山水水、月光竹林的隐士意境中，找到心灵的平衡和安宁。

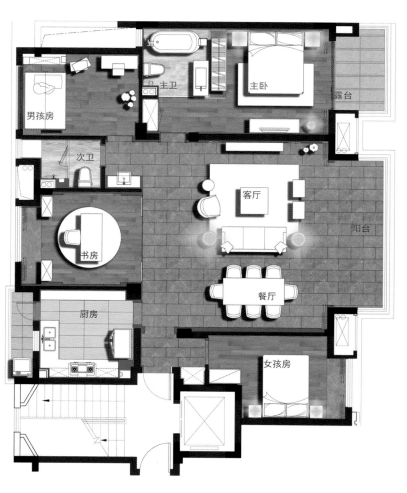

>>> 女孩房：蝴蝶蹁跹

以小女孩的爱好出发，将围棋元素注入空间，选择纯洁的白为主色调。娉婷起舞的彩蝶挂画、犹带生机与温暖的花艺摆件以清秀低调的步子在简洁的空间中娓娓道来，共同营造了一个充满童真且温馨的小世界。

午后伯爵茶
AFTERNOON EARL GRAY TEA

>>> 设计理念 | DESIGN CONCEPT

Freshness of the bergamot wakes up the Monday morning; sweet flores aurantii brings laziness of Wednesday afternoon; in the night with rosemary, stars twinkle; the story of fragrance, is her joys and sorrows.

The starry night in the Rhone River flashes into her heart; almond in blossom reveal the innermost feelings; the silk scarf is the iris in the vase; wearing it, she has eye affinity with Mr. Van Gogh.

Taking a sip of earl gray tea, baking under the sun lazily, then eating a breath mint, wandering in the day and secretly kissing the cheek of night, you will be infatuated with natural sceneries and enjoy a good time with your family.

　　佛手柑的清新，唤醒周一的清晨，甜美的橙花，带出周三午后的慵懒，迷迭香的夜晚星辰闪烁，香味的故事，就是她的喜怒哀乐。

　　罗纳河上的星夜闪进心房，看花季的巴旦杏倾诉衷肠，丝巾是花瓶中的鸢尾，围在颈上，她跟梵高先生总是很有眼缘。

　　呷一口伯爵茶懒洋洋地晒太阳，再吃一颗薄荷糖，在白昼中徜徉，偷偷亲吻夜的脸庞，只愿迷恋风花雪月事，纵享家人相聚好时光。

项目名称：南昌绿地朝阳中心精装户型样板间
软装设计：成象软装
空间设计：梁志天设计师有限公司（SLD）
项目地点：江西南昌

主体色
PRIMARY COLOR

点缀色
INTERSPERSED COLOR

>>> 色彩分析 ｜ COLOR ANALYSIS

Life is art for her which needs to be carefully spent while art itself is her daily life. Considerate and unique silver gray background color is neither flamboyant nor presumptuous, only brings comfort and fashion. Relaxing and leisure tonality makes the space fully play the role of private enjoyment. Simple white and noble purple make the colors more transparent and fresh and make the entire heart calm down to add life a simple freedom.

生活对她来说是要认真消遣的艺术，而艺术本身则是她的每日生活。体贴独特的银灰背景色，不张扬不喧宾夺主，只带来舒适与时尚。放松与惬意的格调，让空间充分发挥私享的作用。简单的白、高贵的紫，色彩更加通透鲜明，将整颗心沉静下来为生活增添一份简单的随性。

>>> 客厅：邂逅一场优雅

光的奏鸣曲打造流光溢彩的繁华，即使最深的夜也无法淹没灰调的优雅。晶亮的反射显得既时髦又前卫，镜中花影更添让生活更加美好的灵感。午后的惺忪感，被阳光照耀的全身酥软，一杯伯爵茶，消遣繁忙之余的闲暇时光。

>>> 餐厅：花开人生

处处都有花香，拯救少女心的柔软小生活才是花艺师的生活常态。美好的生活，总伴着花开而来，每一餐的品味之旅如同走在铺满各种鲜花和鲜果的苏格兰小坡上。

>> 书房：光影青春

从明亮的落地窗前俯瞰城市中的车水马龙，自有幽幽的风情。日光轻描淡写，便用光影在宁静的灰墙上镂刻成一缕心头的暖。

>> 主卧：柔情酣畅

暗香浮动恰好，借着袅娜的光线，倾诉主人的柔情蜜意。春日枕席懒，闻香好入梦，室中熏起的花草香，淡淡的紫，浓浓的情，让这一觉酣畅舒适。

>> 儿童房：造梦场

旭日微光的童趣，是美的觉醒，如同精灵的世界。清淡和缓的奶白色调带来甜而不腻的童年美梦。

图书在版编目（ＣＩＰ）数据

绘心绘色：家居配色新意象 / 深圳视界文化传播有限公司编． -- 北京：中国林业出版社，2017.5
　　ISBN 978-7-5038-8994-3

Ⅰ．①绘… Ⅱ．①深… Ⅲ．①住宅－室内装饰设计－配色 Ⅳ．① TU241

中国版本图书馆 CIP 数据核字（2017）第 096031 号

编委会成员名单
策划制作：深圳视界文化传播有限公司（www.dvip-sz.com）
总 策 划：万绍东
编　　辑：杨珍琼
装帧设计：黄爱莹
联系电话：0755-82834960

中国林业出版社　·　建筑分社
策　　划：纪　亮
责任编辑：纪　亮　王思源

出版：中国林业出版社
（100009 北京西城区德内大街刘海胡同 7 号）
http://lycb.forestry.gov.cn/
电话：（010）8314 3518
发行：中国林业出版社
印刷：深圳市雅仕达印务有限公司
版次：2017 年 7 月第 1 版
印次：2017 年 7 月第 1 次
开本：235mm×335mm，1/16
印张：20
字数：300 千字
定价：428.00 元 (USD 86.00)